Human Chromosome Variation: Heteromorphism and Polymorphism

Herman E. Wyandt • Vijay S. Tonk

Human Chromosome Variation: Heteromorphism and Polymorphism

Springer

Dr. Herman E. Wyandt
Boston University School of Medicine
Center for Human Genetics
Boston, Massachusetts
USA

and

Acupath Laboratories, Inc.
28 South Terminal Drive
Plainview, New York 11803
USA
hwyandt@hotmail.com

Dr. Vijay S. Tonk
Texas Tech University Health Science
Center
Department of Pediatrics
4th Street 3601
Lubbock, Texas 79416
USA
vijay.tonk@ttuhsc.edu

ISBN 978-94-007-0895-2 e-ISBN 978-94-007-0896-9
DOI 10.1007/978-94-007-0896-9
Springer Dordrecht Heidelberg London New York

Library of Congress Control Number: 2011930400

Printed on acid-free paper

Springer is part of Springer Science+Business Media (www.springer.com)

We dedicate this work our families, to Linda,
to Sunita and Sahil, and to Sachi who will
not be forgotten.

Foreword

Clinical experience in virtually all fields of diagnostic medicine confirms the not infrequent enigma of distinguishing an observation of suspected pathology from normal variation. This inherent difficulty becomes even more pronounced by increasing the depth of diagnostic inquiry, exemplified by the stepwise progression from light microscopy, to electron microscopy, to high-resolution chromosome study and finally to molecular analysis. In all of these endeavors, accurate diagnosis followed by clinical decision-making depends on delineating normal variation from pathologic change.

For the clinical cytogeneticist to incorrectly certify a result as purely a heteromorphism may have critical consequences, including recurrence of mental retardation, a major congenital malformation, or the birth of another affected child. The obverse is, of course, also seriously problematic. Heteromorphisms depend upon the technique used such as the type of chromatin stain or molecular methodology. While the determination that a heteromorphic change is a normal variation and not clinically significant, other vitally important uses are well known. Use of heterochromatin blocks, satellite or repeat sequence regions, or inversions, have proved valuable in paternity evaluation, forensic investigation, following bone marrow transplantation, linkage analysis, genotyping, and for the diagnosis of uniparental disomy.

The advent of microarrays has brought even greater emphasis on the need to determine normal variation. Mental retardation and congenital malformations are frequently due to structural chromosomal rearrangements. Such rearrangements that are larger than 5–10 Mb in size are detectable by conventional cytogenetic examination. However, clinically significant smaller rearrangements may be detectable by FISH or where possible by genomic microarrays. The difficulty in determining whether a ~5 Mb rearrangement is present or significant, would lead to an array-based evaluation. However, CNVs confounding the effort to distinguish a polymorphism from a clinically significant rearrangement, can be equally challenging. While studies of CNVs are ongoing using platforms with increasing genomic coverage, current estimates indicate that a single individual has over 1,000 CNVs. Moreover, many and probably the vast number of the CNVs are inherited. The logistic and economic issue that flows from that realization is the need to analyze parental samples before concluding about clinical significance. Systematic checking of parental samples has enabled the elucidation of dozens of newly recognized de novo microdeletion/duplication disorders, some of which are emerging as phenotypically recognizable syndromes. To what degree somatic or germ line mosaicism may complicate continuing studies, remains to be established. Moreover, unanimity has not always been reached for some observed changes. For example, pericentric inversion of the Y chromosome with breakpoints at p11.2 and q11.2 is not considered by some as a heteromorphism. An important limitation of microarrays is the inability to recognize balanced alterations (such as translocations and inversions) that influence risks of abnormality in future or present offspring.

Notwithstanding the demonstration of an inherited or novel CNV, the question of clinical importance may remain hard or impossible to answer, despite a careful assessment of genes within or proximate to the CNV. This consideration is made more difficult given that the CNVs have an important role in genetic susceptibility for many different genetic disorders.

The authors have thoroughly reviewed what is known about human heteromorphisms and presented a succinct and authoritative guide that will surely influence diagnostic cytogenetic reporting for years to come. It would be wise for every clinical cytogeneticist engaged in diagnostic or research studies to have this reference work at hand to assist in the critical distinction between a benign variant and a pathologic chromosomal rearrangement.

Boston University School of Medicine
Boston, USA

Aubrey Milunsky

Preface to Human Chromosome Variation: Heteromorphism and Polymorphism

In the Atlas of Human Chromosome Heteromorphisms, we emphasized the rapid change in standards of care in clinical cytogenetics—"that today's research almost immediately becomes tomorrow's clinical test. What was once unsolvable becomes approachable with new technologies, almost before the … clinician or laboratory director may be aware they are available." The Atlas did not provide a panacea for such problems, nor does the present volume. It did not distinguish between chromosome variants that are clinically significant and those that are not. The present volume, likewise, falls short of such an endeavor. In the Atlas, we spoke of the problematic nature of heteromorphic regions of the human karyotype, and of the necessity of performing parental studies whenever a striking variant was observed. This approach needs to be emphasized even more strongly for new technologies that are revealing ever more detailed knowledge of variable regions throughout the genome.

Standard methods of identifying most human chromosome abnormalities and variants (heteromorphisms) have been in use for more than four decades. The benign nature of heteromorphism of certain chromosomal regions was established in early population studies and information has not been much improved since. Although laboratories strive for longer chromosomes with higher band resolution, these advancements have not significantly added new variants or aided in interpretation of known variants detectable by standard light microscopy. Fluorescence *in situ* hybridization (FISH) in the 1990s allowed better characterization of some variants and revealed a few new variants that were not detectable by standard cytogenetic methods. Likewise, however, they do not necessarily improve on the distinction between variants that are clinically significant and those that are not.

Improved chip (array) technologies and movement toward large scale personalized DNA sequencing have resulted in the routine detection of large variable copy number DNA sequences (CNVs) that are widely dispersed throughout the human genome. Such DNA sequences (not detectable by standard microscopy) typically flank "hotspots" for duplications and deletions that are associated with genetic disease, many of which are now routinely tested for by high resolution array technology. Whether or not CNVs themselves can be disease causing remains uncertain and may depend on their specific location, but they certainly can result in false-positive or false-negative test interpretations. Hence, a number of CNV data bases have been developed to record information about CNVs that over time will hopeful improve the interpretation of test results.

An additional class of chromosomal variant that was first observed and characterized in the early and mid 1970s, and in which there is now renewed interest, is that of the "common" and "rare" chromosomal fragile sites. Fragile sites on chromosomes have been observed to occur in specific bands, under a variety of *in vitro* conditions, including low folic acid, inhibition of folic acid metabolism, etc. With the exception of the fragile X site at band q27.3, associated with X-linked mental retardation, common fragile sites and most rare fragile sites have no direct clinical association. Common fragile sites can be induced in cultured cells from most people. Rare fragile sites (occurring in less than 5% of people) can also be enhanced in cell culture, some by different conditions than common fragile sites. It is well known that both common and rare fragile sites are sites that are frequently recurrent in chromosome rearrangement, and that there is a strong correlation for a significant number with breakpoints involved in cancer rearrangements. More recent molecular characterization has in fact revealed that several such sites are the co-locations of proto-oncogenes.

In contrast to the previous work entitled "Atlas of Human Chromosome Heteromorphism", the current volume has been titled, "Human Chromosome Variation: Heteromorphism and Polymorphism" to more accurately reflect normal chromosome variants at both the microscopic and submicroscopic levels. While we have retained much of the old "Atlas" as a pictorial representation of common and not so common heteromorphisms, we have eliminated chapters, as wells as material in some chapters that seemed less relevant, while hopefully retaining material that is more applicable. Topics previously treated in separate chapters are now condensed as headings under the general title, "Human Chromosome Methods and Nomenclature", comprising Part I. Part II is an updated pictorial section on chromosome heteromorphisms and FISH variants. At the same time, we have added two new sections (previously not covered in the Atlas): Part III is a review of the common and rare fragile sites, with photographs of many of the most common aphidicolin-induced sites; Part IV is a discussion of polymorphisms and copy number variations (CNVs) involving micro- and minisatellites, oligos and SNPs that cannot be detected except at the molecular level, with references to relevant websites for identifying CNVs.

Acknowledgments

This is a review of the work of many investigators spanning more than five decades of cytogenetic research. It is not possible to adequately represent the early efforts of investigators such as A. Craig-Holms, J.P. Geraedts, P. Jacobs, H. Lubs, W.H. MacKenzie, R.E. Magenis, A.V.N. Mikelsaar, H.J. Müller, S. Patil, P. Pearson, M. Shaw, and many others who perceived the need to study heteromorphisms in populations and who attempted to give order to a complicated topic. We have tried to be thorough in our review but, because of the great volume of literature that has accumulated over time, worldwide, we have inevitably made significant omissions. We regret these oversights and anticipate that our colleagues will inform us of the most serious ones. We also acknowledge the contributions to the literature on the topic by the late Ram S. Verma.

For specific examples of common and rare heteromorphisms, we are grateful for the individual contributions from colleagues around the world. These are acknowledged throughout the book and hopefully will encourage additional contributions of a similar nature in future editions. In this regard, we owe special thanks to Lauren Jenkins at Kaiser Permanente Medical Group (San Jose), who provided us with a significant number of examples of chromosome heteromorphisms without which we may never have started. We must also acknowledge the use of archived material from our respective laboratories. The cytogenetic technologists and associates who helped provide additional examples from these sources include: Xin Li Huang, Alex Dow, Agen Pan, Zhen Kang, Xiao Wu, and Hong Shao in the Center for Human Genetics at Boston University, and Caro E. Gibson, Manju G. Jayawickrama, Eun Jung Lee, Jee Hong Kyhm, Eun-Hee Cho, Pam Nye and Chung-Hwan Yuk in the Cytogenetics Laboratory at Texas Tech University. Sun Han Shim (former post-doctoral fellow), in the Center for Human Genetics, also helped provide key examples of FISH variants. For the remainder, we would be remiss if we did not acknowledge the large amount of published material for which we obtained permission to reproduce in this volume.

This project would not have been completed without the help of our respective departments and support staff. In this regard, we are grateful to the following individuals in the Department of Pediatrics, Texas Tech University Health Sciences Center in Lubbock, TX: Richard M. Lampe, M.D., Chairman; Surendra K. Varma, M.D., Vice Chairman; John A. Berry, Administrator. We are similarly grateful to the following individuals in the Center for Human Genetics, Boston University School of Medicine, Boston, MA: Aubrey Milunsky, MB.B.Ch., D.Sc., Professor of Human Genetics, Pediatrics, Pathology and Obstetrics & Gynecology and Director of the Center for Human Genetics and Jeff Milunsky, M.D., Associate Professor of Pediatrics, Genetics and Genomics and Associate & Clinical Director of the Center for Human Genetics.

We owe specials thanks for critical review of the manuscript to: Golder Wilson, M.D, Professor of Pediatric Genetics at Texas Tech University Medical Center, and to Caro Gibson for screening of the manuscript for clerical errors.

We must also thank Peter Butler, Marlies Vlot, and staff at Springer Science+Business Media for taking a personal interest, both in the Atlas of Human Chromosome Heteromorphisms (Kluwer Academic Publishers) and in present version of Human Chromosome Variation: Heteromorphism and Polymorphism. We are also grateful to Max Haring, publishing editor at Springer SBM/Biomedicine for his help with the present volume.

Contents

Figure Contributors

Arturo Anguiano, M.D. (c17, c29) Quest Diangnostics Incorporated, San Juan Capistrano, CA, USA

Petr Balicek, M.D. (c16) Division of Medical Genetics, University Hospital, Kraklove, Czech Republic

Peter A Benn, Ph.D. (c11, c1, c 41) Division of Human Genetics, University of Connecticut Health Center, Farmington, CT, USA

Center for Human Genetics, Boston University School of Medicine, Boston, MA, USA

Cytogenetics Laboratory, Division of Medical Genetics, Texas Tech University Health Sciences Center, Lubbock, TX, USA

J.J.M. Engelen, Ph.D. (c40) Department of Molecular Cell Biology and Genetics, Universiteit Maastricht, Maastricht, The Netherlands

James M. Fink, M.D., Ph.D. (c37, c38) Hennepin County Medical Center, Minneapolis, MN, USA

Cheong Kum Foong (c31, c32) Cytogenetic Laboratory, Kandang Kerbau Women's and Children's Hospital, Singapore

Steven L Gerson, Ph.D. (c18) Dianon Systems, Stratford, CT, USA

Patricia N. Howard-Peebles, Ph.D. 323 Wrangler Dr, Fairview, TX, USA

Numbers in parentheses with a "c" prefix represent specific atlas contributions. Some variants that were submitted were not able to be included because of redundancy. Nevertheless, those individuals or institutions are listed, but are not followed by "c" number(s). In other instances, submissions were of published material so that appropriate citations have been made accordingly in the text, figure or plate where used, but have not been given "c" number(s). We encourage continued submission of variants, or useful data, which have not already been included in this volume, for possible inclusion in future editions.

Syed M. Jalal, Ph.D. (c42) Cytogenetics Laboratory, Division of Laboratory Genetics, Department of Laboratory Medicine and Pathology, Mayo Clinic and Mayo Foundation, Rochester, MN, USA

Lauren Jenkins, Ph.D. (c2) Kaiser Permanente Medical Group, San Jose, CA, USA

Rhett P. Ketterling, M.D. Division of Laboratory Genetics, Department of Laboratory Medicine and Pathology, Mayo Clinic and Mayo Foundation, Rochester, MN, USA

Roger V. Lebo, Ph.D. Department of Pathology, Children's Hospital Medical Center of Akron, Akron, OH, USA

James Lespinasse, M.D. (c10) Laboratoire de Cytogenetique, Centre Hospitalier, Chambery cedex, France

Brynn Levy, M.Sc (Med), Ph.D. Departments of Human Genetics and Pediatrics, Mount Sinai School of Medicine, New York, NY, USA

Thomas Lynch, M.D. Anzac House, Rockhampton, Old Australia

R. Ellen Magenis, M.D. Department of Molecular and Medical Genetics and Child Development and Rehabilitation Center, Oregon Health Sciences University, Portland, OR, USA

Jim Malone (c39) Akron Children's Hospital, Akron, OH, USA

Patricia M. Miron, Ph.D. (c7, c8, c9, c33, c34, c35, c36) Brigham and Women's Hospital, Boston, MA, USA

Susan Bennett Olson, Ph.D. Department of Molecular and Medical Genetics, Oregon Health Sciences University, Portland, OR, USA

Emelie H. Ongcapin, M.D. (c12) Department of Pathology, Saint Barnabas Medical Center, Livingston, NJ, USA

Shivanand R. Patil, Ph.D. Department of Pediatrics, University of Iowa College of Medicine, Iowa City, IA, USA

Jennifer Phy, DO Department of Obstetrics and Gynecology, TTHSC, Lubbock, TX, USA

Sayee Rajangam Department of Anatomy, St Johns Medical College, Bangelore, India

Birgitte Roland, M.D. (c30) Department of Histopathology, Foothill Hospital, University of Calgary, Calgary, Canada

Jacqueline Schoumans (c19, c20, c21, c22, c23, c24, c25, c26, c27, c28) Department of Medical Genetics, University Hospital Haukeland, Bergen, Norway

Cathy M. Tuck-Miller (c15) Department of Medical Genetics and Genetics-Birth Defects Center, University of Southern Alabama, Mobile, AL, USA

Gopalrao V.N. Velagaleti, Ph.D. Departments Pathology, University of Texas Health Sciences Center, San Antonio, TX, USA

Peter E. Warburton, Ph.D. Department of Human Genetics, Mount Sinai School of Medicine, New York, NY, USA

Sharon L. Wenger, Ph.D. (c3, c4) Department of Pathology, West Virginia University, Morgantown, USA

Golder N Wilson, M.D. Clinical Professor of Pediatrics, TTUHSC Amerillo & Lubbock, Lubbock, TX, USA

Professor of Obstetrics and Gynecology, TTHSC Amerillo, Kinder Genome Pediatric Genetics, Plano, TX, USA

K Yelavarthi, Ph.D. (c13, c14) West Virginia University, Morgantown, WV, USA

Adriana Zamecnikova (c5, c6) Department of Genetics, National Cancer Institute, Klenova, Slavaki

J. Zunich, M.D. (c13, c14) Northwest Center for Medical Education, Gary, IN, USA

Part I
Human Chromosome Methods and Nomenclature

Chapter 1
Introduction

As stated in the Atlas of Human Chromosome Heteromorphisms [1], "The cornerstone of genetics is variation…". Heteromorphism, defined in the Atlas as a "microscopically visible chromosome region that is variable in size, morphology and staining properties in different individuals", is just one form of normal variation in the human genome. The terms heteromorphism, normal variant and polymorphism are often used interchangeably, and although there are subtle distinctions, all three usually refer to variations in the human karyotype or genome that are heritable.

Heteromorphisms that are detectable at the microscope level by various banding techniques have been used in a variety of ways. "Because heteromorphisms are stable, highly variable and inherited in a predictable fashion, they [have been historically] useful [and accurate] tools…" [2]: (1) in paternity studies [3]; (2) in determining the parental and meiotic origins of acrocentric trisomies, especially trisomy 21 [4–9], triploidy [10–13], and *de novo* structural abnormities such as 15q deletions in Prader-Willi/Angelman syndromes [14–17]; (3) in determining the mechanisms of origin of various marker chromosomes [17]; (4) in studying uniparental disomy [2]; (5) in establishing the maternal origin of ovarian teratomas [18] and the parental origin of partial and complete hydatidiform moles [19]. More recent studies with molecular polymorphisms have confirmed the accuracy of these studies with amazingly similar results [20–21].

The original Atlas also described variants that are detectable by FISH, but not necessarily visible morphologically. These especially include subtypes of satellite DNA, highly repetitive DNA sequences that make up the most strikingly visible heteromorphic regions on chromosomes 1, 9 16, Y and the acrocentric chromosomes, as well as the pericentric regions of every chromosome (see Levi and Warburton, 2004) [22]. Subtelomeric FISH variants were described as a form of euchromatic variant [23], but not always visible cytogenetically. More recent detection of subtelomeric abnormalities by CGH array technology (see Part IV, Chapter 32) reveals a great amount of copy number variation (CNV) categorizing them as both FISH and array variants.

Polymorphism, defined as one of multiple forms of a normal gene or molecule, present in at least 1% of the population [24–25], was largely ignored in the original Atlas. In contrast to heteromorphism, polymorphisms refer mainly to variants that

H. E. Wyandt, V. S. Tonk, *Human Chromosome Variation: Heteromorphism and Polymorphism,* DOI 10.1007/978-94-007-0896-9_1,

are more often detectable at the molecular level. Several types of heritable DNA polymorphism in the human genome include single nucleotide polymorphisms (SNPs), simple deletion or insertion of a single nucleotide sequence, short tandem repeat (5–25 copies) of a 2–4 bp sequence or *microsatellite* (STRP), variable number tandem repeat polymorphism (multiples of 10^2–10^3 copies) of a 10–100 bp sequence or *minisatellite* (VNTR), and CNP or CNV (two or more copies) of a large 200 bp–1.5 Mb sequence (CNP or CNV) [25]. In particular CNP or CNV has moved to the forefront as a significant form of polymorphism that, when detected as a microduplication or deletion by microarray technology, may easily be misinterpreted as being clinically significant. Numerous data bases are now available to identify literally thousands of CNVs distributed throughout the human genome.

A third type of variant that was touched upon, but was not discussed in detail in the Atlas is the "fragile site". Fragile sites are visible under the microscope, but only under special culture conditions. They can be classified into two basic types, "rare" and "common". While both require special culture conditions to be expressed, rare sites show up only in certain kindreds and hence are regarded to be heritable variants, whereas common fragile sites can be observed in everyone's cells under the right culture conditions. In themselves, common sites are regarded as innocuous; however numerous reports recognize such sites as prone to recurrent rearrangement. More recent investigation has shown several of them to incorporate genes that are frequently rearranged or mutated in cancer. In addition to heteromorphisms and FISH variants, the present volume has added two new sections, one devoted to fragile sites and their properties, and one on CNVs.

References

1. Wyandt HE, Tonk VS (eds) (2004) Atlas of human chromosome heteromorphisms. Kluwer, Dordrecht
2. Olson SB, Magenis RE (2004) Technical variables and the use of heteromorphisms in the study of human chromosomes. In: Wyandt HE, Tonk VS (eds) Atlas of human chromosome heteromorphisms. Kluwer, Dordrecht, pp 63–73
3. Olson SB, Magenis RE, Lovrien EW (1986) Human chromosome variation: the discriminatory power of Q-band heteromorphism (variant) analysis in distinguishing between individuals, with specific application to cases of questionable paternity. Am J Hum Genet 38:235–252
4. Robinson JA (1973) Origin of extra chromosome in trisomy 21. Lancet 1:131–133
5. Schmidt R, Dar H, Nitowsky HM (1975) Origin of extra 21 chromosome in patients with Down syndrome. Ped Res 9:367a
6. Wagenbichler P (1976) Origin of the supernumerary chromosome in Down's syndrome. ICS 397. V International Congress Hum Genet. Chicago, Excerpta Medica, p 167a
7. Magenis RE, Overton KM, Chamberlin J, Brady T, Lovrien E (1977) Parental origin of the extra chromosome in Down's syndrome. Hum Genet 37:7–16
8. Mikkelsen M, Poulsen H, Grinsted J, Lange A (1980) Non-disjunction in trisomy 21: Study of chromosomal heteromorphisms in 110 families. Ann Hum Genet 44:17–28
9. Magenis RE, Chamberlin J (1981) Parental origin of nondisjunction. In: De la Cruz FF, Gerald PS (eds) Trisomy 21 (Down Syndrome): research perspectives. University Press: Baltimore

10. Jonasson J, Therkelsen AJ, Lauritsen JG, Lindsten J (1972) Origin of triploidy in human abortuses. Hereditas 71:168–171
11. Kajii T, Niikawa N (1977) Origin of triploidy and tetraploidy in man: 11 cases with chromosome markers. Cytogenet Cell Genet 18:109–125
12. Jacobs PA, Angell RR, Buchanan IM, Hassold TJ, Matsuyama AM, Manuel B (1978) The origin of human triploids. Ann Hum Genet 42:49–57
13. Lauritsen JG, Bolund L, Friedrich U, Therkelsen AJ (1979) Origin of triploidy in spontaneous abortuses. Ann Hum Genet 43:1–5
14. Olson SB, Magenis RE (1988) Preferential paternal origin of *de novo* structural chromosome rearrangements. In: Daniel A (ed) The cytogenetics of mammalian autosomal rearrangements. Liss, New York, pp 583–599
15. Butler MG, Palmer CG (1983) Parental origin of chromosome 15 deletion in Prader-Willi syndrome. Lancet 1:1285–1286
16. Magenis RE, Toth-Fejel S, Allen LJ, Black M, Brown MG, Budden RC, Friedman JM, Kalousek D, Zonana J, Lacy D, LaFranchi S, Lahr M, Macfarlane J, Williams CPS (1990) Comparison of the 15q deletions in Prader-Willi and Angelman syndromes: specific regions, extent of deletions, parental origin, and clinical consequences. Am J Med Genet 35:333–349
17. Maraschio P, Zuffardi O, Bernardi F, Bozzola M, DePaoli C, Fonatsch C, Flatz SD, Ghersini L, Gimelli G, Loi M, Lorini R, Peretti D, Poloni L, Tonetti D, Vanni R, Zamboni G (1981) Preferential maternal derivation in inv dup(15): analysis of eight new cases. Hum Genet 57:345–350
18. Linder D, McCaw BK, Hecht F (1975) Parthenogenic origin of benign ovarian teratomas. N Engl J Med 292:63–66
19. Kajii T, Ohama K (1977) Androgenetic origin of hydatidiform mole. Nature 268:633–634
20. Knoll JHM, Nicholls RD, Magenis RE, Graham JM Jr, Lalande M, Latt SA (1989) Angelman and Prader-Willi syndromes share a common chromosome deletion but differ in parental origin of the deletion. Am J Med Genet 32:285–290
21. Nicholls RD, Knoll JH, Glatt K, Hersh JH, Brewster TD, Graham JM Jr, Wurster-Hill D, Wharton R, Latt SA (1989) Restriction fragment length polymorphism with proximal 15q and their use in molecular cytogenetics and the Prader-Willi syndrome. Am J Med Genet 33:66–77
22. Levi B, Warburton P (2004) Molecular dissection of heteromorphic regions. In: Wyandt HE, Tonk VS (eds) Atlas of human chromosome heteromorphisms. Kluwer, Dordrecht
23. Jalal SM, Ketterling RP (2004) Euchromatic variants. In: Wyandt HE, Tonk VS (eds) Atlas of human chromosome heteromorphisms. Kluwer, Dordrecht
24. Vogel F, Motulsky AG (1982) Human genetics. Springer, Berlin, p 373
25. Nussbaum RL, McInnes RR, Willard HF (2007) Thompson and Thompson genetics in medicine. Saunders-Elsevier, Philadelphia

Chapter 2
Chromosome Heteromorphism

The term heteromorphism is especially applicable to normal variants observed by chromosome banding techniques. However, normal variations in morphology in certain regions of the human genome were noted even before the advent of chromosome banding. In the first Conference on Standardization in Human Cytogenetics in Denver in 1960 [1], chromosomes were divided into Groups A-G based on their relative sizes and positions of the centromeres. The X chromosome fell somewhere in the C-group. The Y was distinguishable from the G-group by its lack of satellites and somewhat distinctive morphology. At the London Conference in 1963 [2], prominent secondary constrictions were identified near the centromeres in the no. 1 chromosome pair in the A- group, in a chromosome pair (no. 9) in the C-group and in a pair (no. 16) in the E-group. By the Chicago Conference in 1966 [3], it was generally recognized that these regions and the Y varied in length, and that there were morphological variations in the short arms of the D- and G-group chromosomes.

In the early 1970s, Q-, G- and C-banding techniques became widely used. Q- and G-banding introduced a new era in which individual chromosomes could be definitively identified. With this capability, it also became possible to localize regions variable in size and staining to specific chromosomes. In particular, Q- and C-banding revealed distinct classes of heteromorphisms that were not necessarily detectable in non-banded chromosomes, but could be shown to be heritable in banded chromosomes. The most distinctive heteromorphism by Q-banding was the brightly fluorescent distal long arm of the Y chromosome. The size of this brightly fluorescent segment varied from being almost negligible in size to being the longest segment on the Y long arm. Q-banding (Fig. 2.1) also revealed variations in staining of chromosomes 3, 4, 13–15, and 21–22 [4–8].

Although G-banding techniques became widely used for chromosome identification (Fig. 2.2), C-banding revealed size variations of heterochromatin (h) around the centromeres of every chromosome that could be more easily quantitated than in non-banded chromosomes. The h regions of chromosomes 1, 9, 16 and in the distal long arm of the Y, evident in non-banded chromosomes, were especially visible by C-banding (Fig. 2.3) [8–12]. A system to describe variations observed by Q- and C-banding by intensity and size was incorporated into the cytogenetic nomenclature (Table 2.1).

H. E. Wyandt, V. S. Tonk, *Human Chromosome Variation: Heteromorphism and Polymorphism*, DOI 10.1007/978-94-007-0896-9_2,

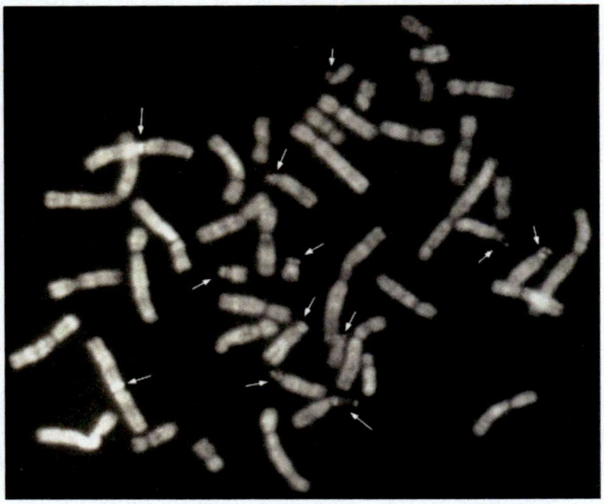

Fig. 2.1 Normal Q-banded metaphase from 46,XX, female showing heritable variations in size and intensity of staining (*arrows*), especially of centromeric region of chromosome 3s, and of centromere, short arm, stalk and satellite regions of acrocentric chromosomes

Additional specialized techniques quickly followed, including R-banding [13], silver staining for nucleolar organizing regions (NORs) [14], G-11 staining [15], and staining with various fluorescent DNA-binding fluorochromes, either singly or in combination. Some of these techniques revealed additional subclasses of variants [16] so that a complex system of characterizing variants by band intensity and stain-

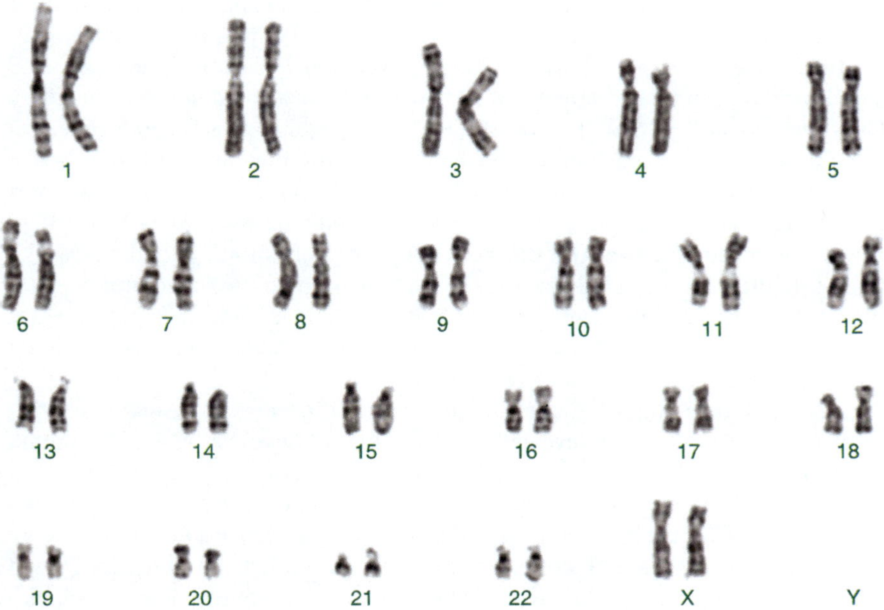

Fig. 2.2 Normal female karyotype by GTG banding

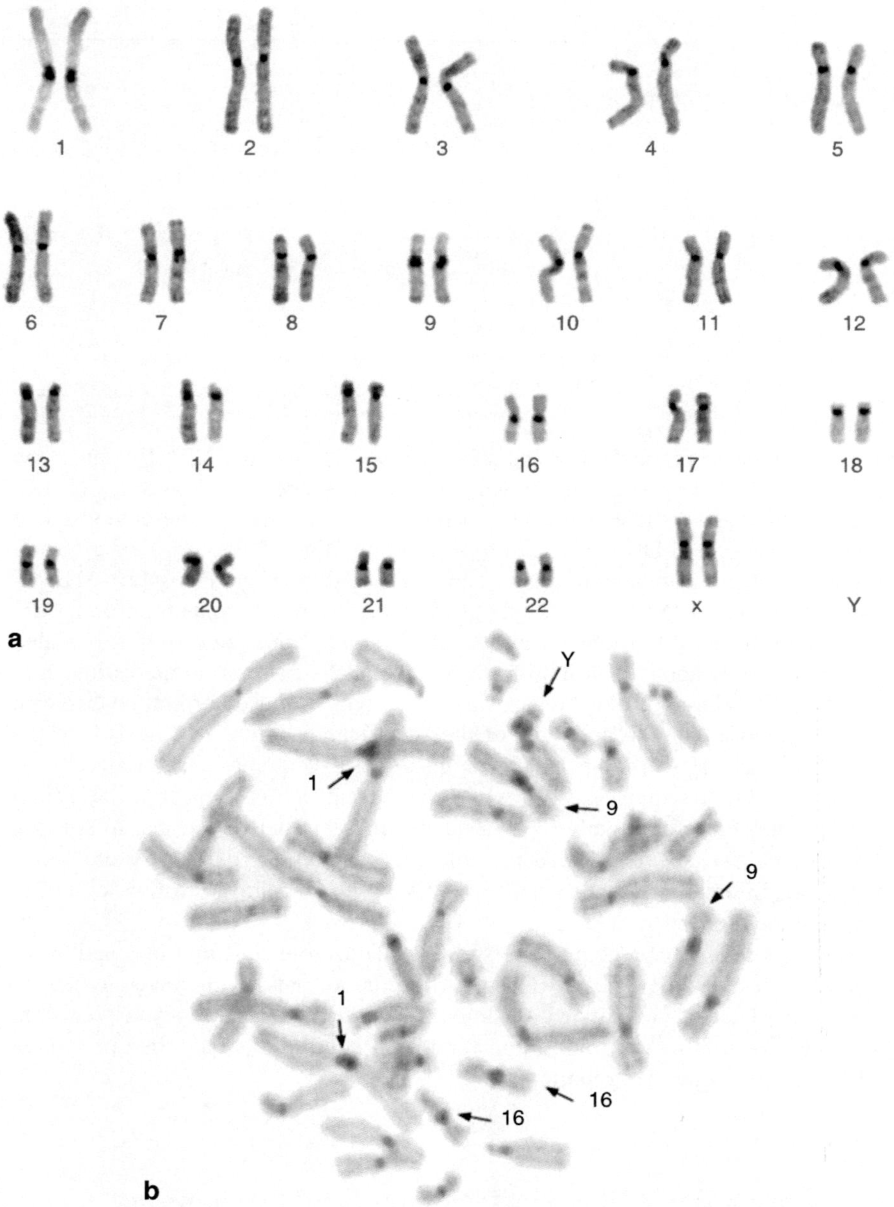

Fig. 2.3 **a** Normal female karyotype by CBG banding by barium hydroxide treatment followed by Giemsa staining (c41). **b** CBG-banded metaphase from normal male. Arrows point to 1, 9, 16 and Y chromosomes, which typically show the greatest amount of heteromorphism in different individuals

Table 2.1 Numerical expression of intensity of Q and size of C bands. (Adapted from Paris Conference, 1971 [16])

Size	Intensity
Q-banding	
1 Very small	1 Negative (no or almost no fluorescence)
2 Small	2 Pale (as on distal lp)
3 Intermediate	3 Medium (as the two broad bands on 9q)
4 Large	4 Intense (as the distal half of 13q)
5 Very large	5 Brilliant (as on distal Yq)
C-banding	
1 Very small	0 No quantitation of intensity
2 Small	
3 Intermediate	
4 Large	
5 Very large	

ing technique was proposed in a Paris Conference Supplement [17]. However, the system was not widely used, and is not included in subsequent versions of ISCN [18].

Early molecular studies showed C-band heteromorphisms to be composed of different fractions of DNA, referred to as satellite DNAs based on their differing AT/GC content and buoyant densities in CsCl or Cs_2SO_4 gradients [19–21]. Alkaline Giemsa and DA/DAPI [22, 23] techniques stain components of 1qh, 9qh, D-G-group short arms, 16qh and distal Yqh. *In situ* hybridization studies revealed different but overlapping distributions of satellite DNA fractions to the various heterochromatic regions in the human karyotype, with a loose correlation between alkaline Giemsa staining and sites of the "classical" satellite III [24–28]. In the 1980s and early 1990s, molecular techniques more accurately characterized various satellite DNA sequences [29], while fluorescent *in situ* hybridization (FISH) [30, 31] allowed virtually any DNA sequence to be visually localized to specific chromosomal sites. Current FISH and molecular technologies define satellite DNAs somewhat differently, but sequences in the satellite III family localize to similar chromosome regions.

FISH and DNA sequencing have shown considerable shuffling of satellite sequences. These technologies provided the means not only to characterize heteromorphisms detected by classical techniques with greater accuracy and precision, but to also identify new chromosomal variants. A handful of what might be termed "FISH variants" has been reported.

2.1 Chromosome Banding Techniques and Mechanisms

2.1.1 Q-banding

Caspersson and colleagues at the Karolinska Institute with an American team of biochemists at Harvard Medical School headed by S. Farber and G. Foley set out to test or design fluorescent molecules that would preferentially bind to specific nucleotide pairs in DNA, which they hoped to be able to detect spectrophotometrically. One molecule tested was quinacrine mustard dihydrochloride (QM), a nitrogen mustard analog of the anti-malarial drug, quinacrine. The dye, first applied to *Vicia faba* and *Trillium erectum*, revealed brightly fluorescent bands that distinguished the individual plant chromosomes. The findings led Caspersson et al. [32] to apply QM staining to human chromosomes X, with the discovery that the end of the long arm of the Y chromosome was brightly fluorescent -bright enough that the human Y chromosome could be easily detected in interphase as well as in metaphase cells. With refinements, QM produced banding patterns that were specific for each human chromosome and revealed heritable variations in size and/or intensity of certain regions, especially of distal Yq, of the centromeric regions of chromosomes 3 and 4, and of the centromeric and short arm regions of the acrocentric chromosomes (Fig. 2.1). Several investigators showed that the AT-rich regions of DNA corresponded to the bright fluorescent bands obtained with quinacrine mustard [33–35]. Weisblum and DeHaseth [33] showed that rather than preferential binding, this difference in intensity of fluorescence reflected a difference in quenching of the QM molecule. AT-richness alone, however, is not the sole determinant of the intensity of Q-banding. The actual differences in relative percentages of AT vs. GC in different regions are not as great as might be implied. The periodicity of interspersed GC, within short, highly repetitive AT-rich sequences, as well as the presence of specific nucleoproteins appears to play a significant role [36].

2.1.2 G-banding

G-banding, introduced in 1971 by Sumner et al. [37] overcame two significant problems of Q-banding (stability and cost) and thus became the more widely used banding technique in clinical laboratories. G-banding acronyms GTG, GTW, GTL and GAG all represent variations used to obtain the same banding pattern that can be seen and analyzed by standard light microscopy. While the original G-banding method used acid fixation with saline treatment followed by Giemsa staining (GAG) [38], application of proteolytic enzymes such as trypsin [39, 40] or pancreatin [41, 42] were simpler and improved the banding pattern. The blood stains, Wright's or Leishman's, are often used instead of Giemsa, depending on the laboratory's experience and preference. G-banding patterns are identical, however, irrespective of how they are obtained (by enzymatic or chemical pretreatment) or the blood stain used (Fig. 2.2). Similar to bright Q-bands, dark-staining G-bands are AT-rich regions of chromosomal DNA that are more condensed, and replicate their DNA later than GC-rich regions which are less condensed (Table 2.2) [43]. DNA-binding proteins thought to be involved in maintaining chromosomal structural integrity form the nuclear matrix and include topoisomerases that have a basic role in the control of gene activity [44–46]. It may be that nuclear matrix proteins that hold

Table 2.2 Techniques for recognition of different classes of chromatin and properties of chromosome bands. (Modified from Sumner [43])

Class			Properties of chromosome bands	
Heterochromatin	Euchromatin	Special regions	Positive G-/Q-bands, negative R-bands, chromomeres (pachytene)	Negative G-/Q-bands, positive R-bands, interchromomeres (pachytene)
• C-banding • G-11 banding • Q-banding • Distamycin/DAPI	• G-banding • Q-banding • R-banding • T-banding • Replication banding	• C-banding • Cd-banding • Immunofluorescent staining withCREST serum	• Early chromatin condensation • Late DNA replication • AT-rich DNA • Tissue-specific genes • Long intermediate repetitive sequences (LINEs)	• Late chromatin condensation • Early DNA replication • GC-rich DNA • Housekeeping genes • Short intermediate repetitive sequences (SINEs)

AT rich regions together make them less easily available for DNA replication and at the same time allow dye to bind only in monomer form so that they stain more intensely. Conversely, GC-rich regions that are gene-rich and transcriptionally active may be more loosely bound and consequently bind dye in polymer form with less intense staining. Giemsa, Leischman, Wright or Romanowski blood stains all contain mixtures of thiazin dyes, each of which can produce banding under the right conditions. It is evident from the variety of treatments that produce G-banding that more than one mechanism is involved. The most reliable and widely used treatment is mild proteolytic digestion with trypsin [39, 40]. However, the precise role of nucleoproteins in G-banding has not been determined [47–50]. Extraction of histones also seems to have little effect [51–54]. In fact, very little protein is lost from chromosomes in various G-banding treatments [48]. Furthermore, it is evident that there is an underlying structural integrity of the chromosome that is revealed in the "chromomere pattern" of very long chromosomes in meiosis [55, 56]. This pattern in non-banded meiotic chromosomes is identical to the pattern of G-banded metaphase chromosomes (see ISCN [18]).

The relationship between DNA structure and the binding of components making up Giemsa dye mixtures also is not totally understood. Treatments that loosen the integrity of underlying DNA structure appear to be most effective, suggesting that certain Giemsa components bind to condensed DNA in monomeric form and to looser DNA structure in polymeric form. The more the individual dyes components become stacked, the greater the shift to lower absorption spectra (purple or pink). In monomer form the shift is to the blue end of the spectrum. Such a shift in color, based on a dye's ability to become stacked in polymer form, is referred to as metachromacy. Some Giemsa components are more metachromatic than others. Methylene blue, Azure A, Azure B, and Thiazin show varying degrees of metachromacy determined by the number of methyl groups present on the dye molecule [57–59]. Eosin, which is also a component of Giemsa dyes, shows no metachromacy but appears to have a differential staining effect when combined with the other components.

2.1.3 R-banding

Utrillaux and Lejeune [60] introduced a banding technique involving treatment of chromosomes in saline at high temperature (87°C) that resulted in a reverse pattern of G- or Q-bands. They called this "reverse (R) banding" and, since the method involved staining with Giemsa, it is described as a "RHG" banding. R-bands are most useful in identifying abnormalities involving the terminal regions of chromosomes, which are lighter staining by G- and or Q-banding. Alternate methods to produce R-banding use various fluorescent chemicals such as acridine orange and chromomycin A3/methyl green [61–63] (Table 2.2). However, because of technical difficulties or fluorescent requirements, R-banding is still not used in many laboratories.

2.1.4 C-banding

During experiments with *in situ* hybridization of tritium-labeled satellite DNA to mouse chromosomes, Pardue and Gall [64] noted that constitutive heterochromatin at the centromeres of mouse chromosomes stained darker than other chromosomal regions. In 1971, Arrighi and Hsu [65, 66] developed a modified technique in which they applied Giemsa staining to preparations that were first denatured with 0.07 M NaOH and then incubated in two times standard saline concentration (2 × SSC) for several hours. In a more recent modification, Sumner [67] substituted barium hydroxide for sodium hydroxide, producing the same C-banding pattern (CBG-banding) but with less distortion of the chromosome morphology. Both procedures result in intense staining of the heterochromatin around the centromeres, whereas the rest of the chromosome stains pale blue (Fig. 2.3). Arrighi and Hsu initially postulated this differential staining was due to faster re-annealing of repetitive DNA in heterochromatin than in the less repetitive DNA sequences elsewhere. McKenzie and Lubs [68] produced C-banding by simply treating chromosomes with HCl and prolonged incubation in 2 × SSC. Studies by Comings et al. [48] demonstrated considerable extraction of nucleoprotein and DNA from non-heterochromatic regions by various C-banding treatments, while heterochromatic regions were resistant to such extraction. Furthermore, they demonstrated that hybridization of repetitive sequences in solution was not required for enhancement of staining, but in fact those regions reassociated instantaneously when they were removed from the NaOH solution. Subsequent incubation in 2 × SSC extracted additional non-heterochromatic DNA. Since incubation that produces C-banding is done for times ranging from a couple of hours to overnight, it is unlikely that much single stranded DNA remains to bind Giemsa components. Differential staining is more likely due to the greater amount of double stranded DNA remaining in the heterochromatic regions.

2.1.5 Cd Banding

The technique, first described by Eiberg [69], reveals pairs of dots at presumed centromere locations; hence, the term "centromere dots" (Cd). The technique involves the usual hypotonic treatment of chromosomes followed by a series of fixations starting with a 9:1 ratio of methanol: acetic acid followed by a 5:1 ratio and then the standard 3:1 ratio. One week old slides are then incubated in Earle's balanced salt solution (pH 8.5–9.0) at 85°C for 45 min followed by staining in a dilute solution of phosphate-buffered Giemsa (0.0033 M, pH 6.5). The technique appears to specifically stain only active centromeric regions and not inactive centromeres, secondary constrictions or other variable heteromorphic regions [70, 71]. It has been used identify the active centromere(s) in dicentric, pseudodicentric and Robertsonian translocations. The mechanism of this technique suggested by Eiberg was that it represented a specific DNA-protein complex. Evans and Ross [72] suggested the Cd-positive regions represent kinetochores. Nakagome et al. [70, 71] and Maraschio et al. [73] studied dicentric and pseudodicentric chromosomes and showed that the Cd-positive regions do appear to correspond only to active centromeres. The presence or absence of specific centromeric proteins associated with centromeric activity have been recently studied with specific fluorescent antibodies that distinguish particular proteins associated with active or inactive centromeres [74].

2.1.6 G-11 Banding

G-11 staining is used to selectively stain some heterochromatic regions on human chromosomes a deep magenta color in contrast to the pale blue color of the remainder of the chromosome. These include chromosomes 1, 3, 5, 7, 9, 10, 19 and Y. However, there is variability in the intensity of staining at the pericentromeric and satellite regions of acrocentric chromosomes. Such variability is dependent on the individual characteristics of these chromosomes. The G-11 technique utilizes modified Giemsa staining at an alkaline pH [75] and is useful in the study of human heteromorphic variants and pericentromeric inversions, especially on chromosome 9. Figure 2.4a shows a metaphase with typical G-11 banding. G-11 banding received its name from attempts to obtain differential banding of specific chromosome regions by staining in Giemsa at different pH values. The standard pH of the staining solution in G-banding procedures is 6.8–7.0. It was found by Patil et al. [77] that if the alkalinity of some Giemsa mixtures was raised to 9.0, G-banding could be achieved without any other special treatment. Bobrow et al. [76] showed that if alkalinity was raised to pH 11, subcomponents of C-bands, especially the secondary

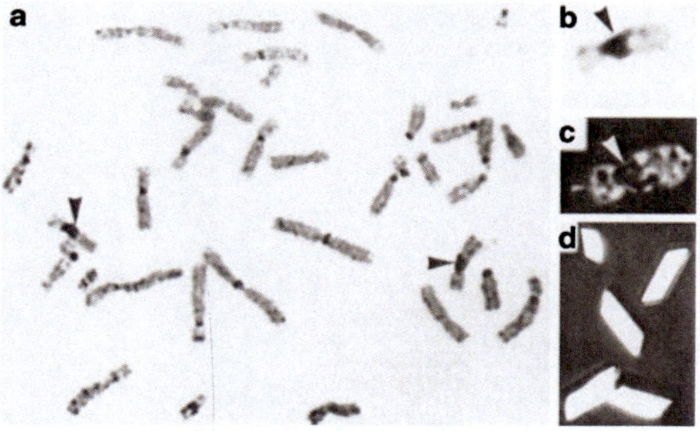

Fig. 2.4 **a** Metaphase showing G-11 banding with inset, **b** showing enlarged 9 by bright-field and inset, **c** showing same chromosome 9 by phase contrast microscopy. Inset **d** shows large trapezoidal azure-eosinate crystals by phase contrast microscopy. [Modified from Wyandt et al. (1976) Exp Cell Res 102:85–94]

constriction (qh region) of chromosome 9 stained a deep magenta color in contrast to the pale blue color of the euchromatic regions. Jones et al. [20] first showed that satellite III DNA, isolated on a silver cesium sulfate gradient, hybridized to the heterochromatic regions of chromosome 9 and to the acrocentric chromosomes. Buhler et al. [28] showed that this magenta-staining DNA which appears to be especially specific for 9qh, 15p and Yq corresponded to sites of hybridization of a specific class of highly repetitive satellite III DNA. Other classes of satellite DNAs, I-VII were found to be distributed in chromosome 9 and in other chromosomes [21], but satellite III was found mainly in these three chromosomes. The mechanism of G-ll banding is still uncertain. Wyandt et al. [77] tested various components of Giemsa and showed that G-11 banding could be achieved when the right proportions of Azure B and Eosin Y were mixed at pH 11. When mixed in equimolar amounts, most of the Azure B and Eosin Y precipitated as large highly reflective trapezoidal crystals of azure-eosinate (Fig. 2.4d). Finer crystals appear to be precipitated at magenta colored sites on chromosomes (Fig. 2.4b, c).

2.1.7 Silver Staining (AgNOR)

Silver staining is a method to stain the nucleolar organizer regions (NORs) on the human acrocentric chromosomes. NORs, which contain the genes for ribosomal RNA or proteins, were known early to stain with silver. Using this information, Howell et al. [78] showed that NORs on chromosomes could be stained with silver nitrate and called their technique "Ag-SAT". Howell and Black [79] subsequently developed a simplified technique using a colloidal developer for better results. Many laboratories use this method with various modifications. Figure 2.5 shows a metaphase with typical AgNOR staining.There is still controversy as to the nature or exact location of this silver staining. Miller et al. [80] showed that the activity of NOR regions appeared to be responsible for the staining. Goodpasture et al. [81] showed the actual location of the staining was in the satellite stalks of acrocentric

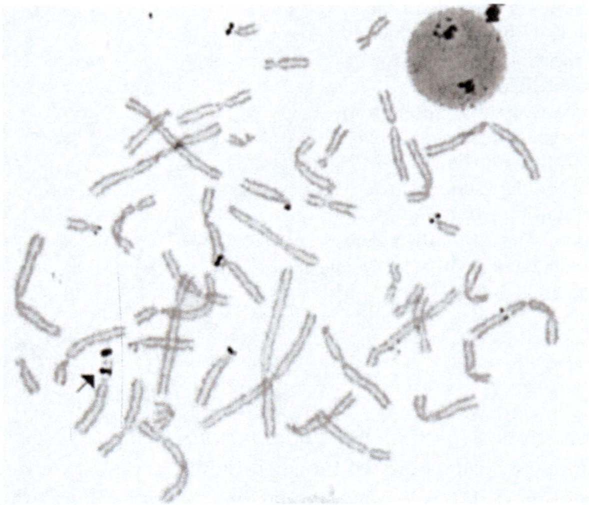

Fig. 2.5 Metaphase showing typical AgNOR staining

chromosomes and not the satellites themselves, although the silver stained mass may appear to cover or extend into the satellite region if the stalk or satellite is small. Subsequent experiments by Verma et al. [82] showed that Ag-NOR positive chromosomes are those that are found frequently in satellite associations while the Ag-NOR negative chromosomes are not seen in such associations. Silver staining is an important banding method to study heteromorphic variations in the size and number of NORs, and to characterize marker chromosomes or other structural rearrangements involving the acrocentric chromosomes.

2.2 Other DNA-Binding Fluorochromes

A variety of different DNA binding fluorochromes will produce chromosome banding patterns or enhancement of AT or GC rich regions depending on absorption and emission spectra and how they are used in combination (Table 2.3). For instance, the combination of distamycin A (DA) and DAPI) produces bright qh regions on chromosome 1, 9 and 16 that correspond to G-11 bands and probably to satellite III DNA. The use of various fluorochromes and their mechanisms of action have been described by others [63] and will not be described in detail here.

2.3 Sister Chromatid Exchange Staining (SCE)

Sister chromatid exchanges (SCE) are the result of interchange of DNA between replication products at homologous loci [83]. SCEs at low levels are normally seen in humans and can be demonstrated in somatic cells by incorporating a thymidine

Table 2.3 Fluorescent DNA ligands used in human chromosome staining, base affinity and type of banding when used with counter stain. (Adapted from Verma and Babu [63])

Primary dye	Affinity	Counter stain	Banding
DAPI	AT	Distamycin A[a]	DAPI/DA
DIPI	AT	Netropsin	DAPI/DA
		Pentamidine	DAPI/DA
Hoechst 33258	AT	Distamycin A[a]	DAPI/DA
		Netropsin	DAPI/DA
		Actinomycin D[b]	QFH-bands
		Chromomycin A_3	QFH-bands
7-aminoactinomycin D	GC	Methyl green[a]	R-bands (enhanced)
Chromomycin A_3	GC	Distamycin A[a]	R-bands (enhanced)
Mithramycin	GC	Malachite green[a]	R-bands (enhanced)
Olivomycin	GC	Distamycin A	R-bands (enhanced)
		Netropsin	R-bands (enhanced)
		Methyl green[a]	R-bands (enhanced)
Coriphosphin		Methyl green	R-bands (enhanced)
Quinacrine/quinacrine mustard	GC (low)		Q-bands

[a] Non-fluorescent with AT affinity
[b] Non-fluorescent with GC affinity

analog, 5-bromodeoxyuridine (BrdU) into replicating DNA for two successive cell cycles and subsequent photodegradation of the resulting chromosomes. Staining of metaphases with Hoechst 33258 [84] or with Giemsa following this procedure results in faint staining of one chromatid and strong staining of the other chromatid. A reversal of staining intensity of the two chromatids occurs where there has been an exchange (Fig. 2.6). Because of the semi-conservative nature of DNA replication, after two complete pulses of BrdU substitution, one chromatid has both halves of

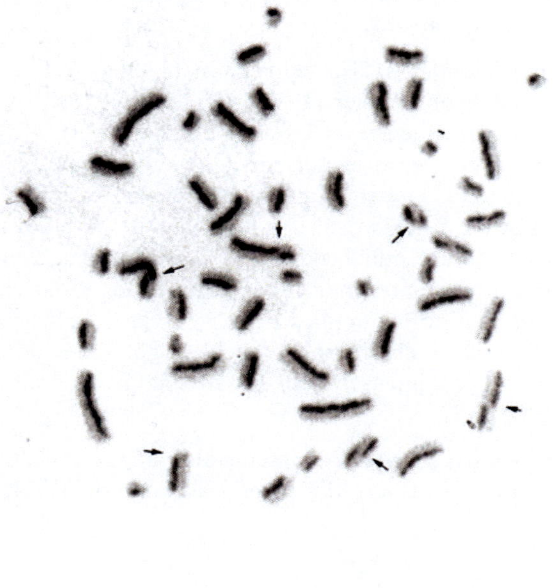

Fig. 2.6 Metaphase showing sister chromoatid exchanges (*arrows*)

the DNA helix BrdU-substituted (bifilarly labeled) while the other chromatid has only one half of the DNA helix BrdU-substituted (monofilarly labeled). The latter is the basis of differences in staining of sister chromatids that allow detection of SCEs, mainly in non-heterochromatic regions. The technique has been extensively used for testing the mutagenic potential of various chemicals [85], to study cell cycle kinetics [86, 87] and to diagnose Bloom syndrome, in which there is a tenfold increase in SCE per cell [88].

2.4 Replication Banding

Replication banding is most useful in identifying the early and late replicating X-chromosomes in females or in patients with sex chromosome abnormalities. It is well known that one of the X-chromosomes in females is inactive, resulting in dosage compensation [89]. It is also known that X chromosome inactivation is random and that the inactive X chromosome initiates and completes DNA synthesis later than the active X and other chromosomes [90–94]. Replication banding, obtained by incorporation of 5-bromodeoxyuridine (BrdU) and subsequent staining with Giemsa or other stains [84], allows distinction of the active and inactive X-chromosomes. Variations in replication banding can also be achieved. In the "T pulse" procedure, BrdU is made available at the beginning of the cell cycle and then replaced with thymidine the for last 5–6 h before the harvest. With the RBG technique (R-bands by BrdU and Giemsa), the active or early replicating chromosome regions that inactive X chromosome, stain dark. The "B pulse" is the opposite. Thymidine, made available at the beginning of the cell cycle, is replaced with BrdU the last 5–6 h before harvest. Subsequent Giemsa staining will result in early-replicating chromosome regions appearing dark because they have incorporated thymidine, while the inactive or late-replicating chromosome regions appear pale due to the BrdU-incorporation.

Banding patterns: The equivalent of Q- and G- or R-banding patterns is achieved depending on whether a B or T pulse is used. If a B-pulse is used, a Q or G-banding pattern is achieved and if a T-pulse is used, an R-banding pattern is achieved. Subtle changes in pattern toward the earliest R-bands or latest G-bands can be achieved by shortening the length of the BrdU pulse. A short T-pulse at the very end of the S-period can produce what are referred to as T-bands (bright or dark bands at the terminal ends of some chromosome arms). These bright bands with a T-pulse also correspond to early replicating, GC-rich regions, whereas dull bands correspond to late-replicating AT-rich regions. The exception to this is the late-replicating X chromosome whose bright bands do not differ in AT: GC content from the less intensely stained bands at the same locations on the early-replicating X (Fig. 2.7).

Lateral asymmetry: An interesting variation of the BrdU labeling technique is the method of detecting lateral asymmetry. The latter is due to an interstrand

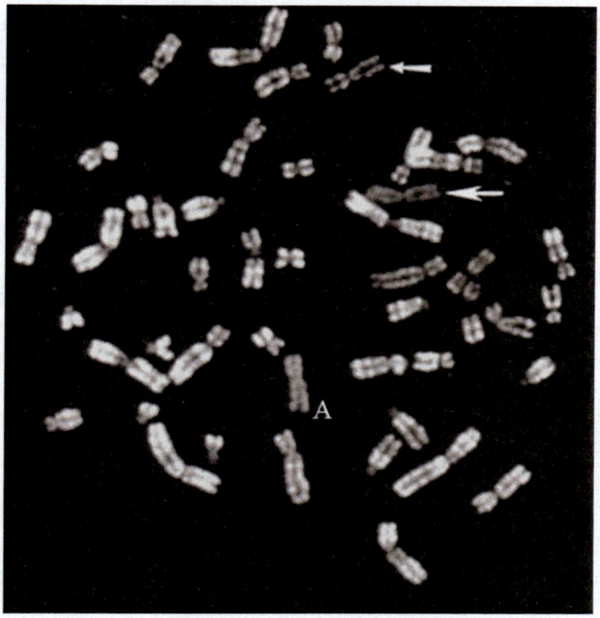

Fig. 2.7 Metaphase with 47,XX,i(Xq) showing replication banding with a late T-pulse showing active X (A), lighter staining inactive X (*small arrow*) and extra i(Xq) (*large arrow*)

compositional bias in which one half of the DNA helix is predominantly T-rich and the complementary half is correspondingly A-rich [95]. Since BrdU substitutes for thymidine and not adenine, after one complete pulse of BrdU, the BrdU-rich strand stains less intensely than the T-rich complement, resulting in a block of heterochromatin that is more intensely stained on one chromatid than on the other. A more equal distribution of thymidine in both strands in either euchromatin or heterochromatin without interstrand compositional bias results in both chromatids staining similarly. Variation in the size and location of such blocks forms the basis of a subclass of variants in chromosomes 1, 9, 15, 16 and Y [96–98].

2.5 High Resolution Banding and Special Treatments

Other treatments and methods that have particular bearing on characterizing heteromorphisms include treatments such as methotrexate added to cultures to synchronize cells in G2 [99] and used for high resolution chromosome banding. Ethidium bromide intercalates into GC rich regions during cell culture, a property that is also used to produce elongated chromosomes for high resolution banding analysis [100, 101]. 5-azocytidine and a number of DNA analogs such as FudR, produce very long secondary constrictions such as shown by Balicek [102] or can enhance so-called "fragile sites" on chromosomes. Most of these are common fragile sites that can be induced *in vitro* in cells from anyone (See Sect. 5 on Fragile Sites). Other "rare" fragile sites are induced only in cells from certain individuals and are heritable.

2.6 Satellite DNA in Heteromorphic Regions

Genes and gene-related sequences (promoters, introns, etc) constitute about 25% of the human haploid genome; only about 3% of the genome is transcribed. Repetitive sequences comprising most of the remainder are the basis both of heteromorphisms observed at the chromosomal level and polymorphisms detected at the molecular level. Tandemly repeated DNA sequences are classified by the length of the individual repeated unit and by total size [103].

Satellite DNA makes up approximately 10% of the genome [104, 105]. Consisting of large tandemly repeated DNA sequences, it is located mainly in heterochromatic blocks in the pericentromeric regions of human chromosomes, the short arms of acrocentric chromosomes and the distal long arm of the human Y chromosome [106–108]. *Alpha satellite* DNA is the principle component found at the centromere of every human chromosome. Other satellite DNAs distributed to various chromosomal locations include: (1) *Beta satellite* DNA, a 68 bp monomer that consists of different subsets that have been shown to be chromosome specific by FISH [106]; (2) *Gamma satellite* DNA, a 220 bp monomer, observed at the centromeres of chromosomes 8 and X [107–110]; (3) additional families that include a 48 bp satellite DNA on the acrocentric chromosomes, and the Sn5 family found in the pericentromeric regions of chromosome 2 and the acrocentric chromosomes [110].

Human satellite DNA fractions, consisting of heterogeneous mixtures of repetitive DNA sequences isolated from main band DNA by buoyant densities on CsCl (cesium chloride) [111] or $CsSO_4$ (cesium sulfate) gradients [21, 24, 111] are referred to as *classical satellites I, II and III* [25, 112]. *In situ* hybridization of these fractions to human chromosomes is to locations that correspond to heterochromatin viewed by C-banding or by the fluorescent dyes DAPI and distamycin. [25–27, 113].

Satellite DNA fractions have been further separated by restriction enzyme analysis into *classical satellites 1, 2 and 3,* found primarily in the large h regions of chromosomes 1, 9, 16 and Y [114–116]. Although, satellites 1, 2 and 3 are incorporated within density gradient fractions, they are distinct from satellites I, II and III [3]. By *in situ* hybridization, satellite 1 is localized to the pericentromeric regions of chromosomes 3 and 4, and the short arms of the acrocentric chromosomes, both proximal and distal to the rDNA of acrocentric stalk regions. Satellite 2 is localized to the large heterochromatic regions of chromosomes 1 and 16, with less prominent domains in the pericentromeric regions of chromosomes 2 and 10. Satellite 3 is localized to h regions of chromosome 1, 9, Y and the acrocentric chromosome short arms, proximal to the ribosomal DNA [114]. It is also found in the pericentromeric region of chromosome 10 [117].

2.6.1 *Alpha Satellite DNA*

The fundamental unit of alpha satellite DNA is a monomer of ~171 bp. Monomers are tandemly organized into higher-order repeats (HORs) ranging from 2 to >30 [118]. HORs at each centromere are in turn tandemly repeated up to several hundred times to form an array of several million base pairs. HORs that are specific for each chromosome and hence useful as FISH probes typically show less than 5% divergence. In addition to HORs, alpha satellite DNA contains a subset of monomers with a 9 bp degenerate motif that serves as the binding site for Centromeric Protein B (CENP-B) found in most mammalian centromeres and initially thought to be involved in recruitment of essential kinetochores proteins such as CENH3 (CENP-A), CENP-C and CENP-E. However, neither alpha satellite motif nor CENP B are required for the formation of a functional centromere in all mammals and more recent investigation indicate epigenetic factors rather than sequence directed mechanisms in the formation of centromeres. The role of alpha satellite motifs and various centromeric proteins in centromere function is still an active area of investigation [119–121].

2.6.2 *Minisatellites*

Levy and Warburton [103] classify minisatellites into AT and GC rich. Tandemly repeated GC rich sequences [122] are at many different loci which vary in the size of the individual repeat (6 to ~100 bp) as well as in total length (100 bp to several kilobases). The widely variable number of tandem repeats (VNTRs) at these loci has made them a useful tool in forensic science for individual identification by DNA fingerprinting, and as highly polymorphic, multiallelic markers for linkage studies [123]. While most minisatellites are GC-rich, AT-rich minisatellites in humans are remarkably different from the GC-rich minisatellites [124–128]. The common features of these alleles include a predicted tendency to form hairpin structures and a domain organization with similar variant repeats commonly existing as blocks within arrays [123]. These loci may also share some mechanisms of mutation, with transient single-stranded DNA forming stable secondary structures which promote inter-strand misalignment and subsequent expansions or contractions in repeat number [128]. Telomeres are a special subset of minisatellites. The majority of hypervariable minisatellite DNA sequences are not transcribed, however some have been shown to cause disease by influencing gene expression, modifying coding sequences within genes and generating fragile sites [123].

2.6.3 *Microsatellites*

Microsatellites consist of units of two to four nucleotides repeated one to a few dozen times. Polymorphic alleles of such sites consist of a differing numbers of repeats, also referred to as short tandem repeat polymorphisms (STRPs). Several hundred thousand STRP loci are distributed throughout the genome with many alleles for each locus in the population. Microsatellite polymorphisms are not usually implicated in disease, but are useful markers for determining the identity of a particular individual [129, 130].

2.7 Single Nucleotide Polymorphisms (SNPs)

The most common polymorphisms are single nucleotide polymorphisms (SNPs). In contrast to STRPs, SNPs usually only have two alleles for any specific location. They occur approximately every 1,000 bp, with approximately 3 million differences between any two genomes or an estimated 10 million alleles in all human populations [131]. A subset of approximately 1 million of the most frequent SNPs have been chosen for a high-density map called the "HapMap" or haplotype map of the human genome [132].

2.8 Fluorescence *In Situ* Hybridization (FISH)

Fluorescence *in situ* hybridization (FISH) has been a powerful adjunct in cytogenetics. In principal, any piece of DNA (or RNA) can be isolated, amplified and labeled. It can then be hybridized to intact chromosomes, nuclei, or other fixed target and detected. Labeled DNA segments, called probes, are prepared by a variety of techniques including (1) synthesis of cDNAs from mRNAs by reverse transcriptase [132]; (2) isolation of specific sequences by PCR amplification and/or gel electrophoresis [133, 134]; (3) propagation of larger DNA fragments in bacteria or yeast by insertion into cloning vectors such as plasmids, phage, cosmids, BACS, or YACS [135, 136]; (4) isolation and cloning of partial or complete DNA libraries from specific chromosome regions or entire chromosomes by microdissection [137, 138] or chromosome sorting [139, 140]. Whatever the source, labeling is usually completed by nick translation or random priming with nucleotides that either have fluorescent label attached directly or combined with a ligand that is recognized by a fluorescent-tagged protein.

Procedures for *in situ* hybridization, described in numerous reviews elsewhere [141, 142], involve a number of precisely controlled steps. The crucial steps are: (1) denaturation of probe and target DNA sequences to single strands; (2) incubation of probe and target under conditions that allow specific association (hybridization) of labeled probe DNA to complementary target sequences; (3) washing away non-hybridized probe; (4) detection of hybridized sequences in target cells. The rate of hybridization of probe in solution to complementary DNA targets bound on the glass slide follows first order kinetics [141]. This rate is dependent upon the labeled probe concentration in solution (number of copies of a specific sequence per unit volume) at given time. If the ratio of labeled probe to unlabeled target is too low, insufficient labeled sequence will anneal to the target to permit subsequent detection. If the ratio is too high, precipitation of probe or non-specific hybridization to imperfect complements may result in false-positive signals. Typical ratios of probe to target DNA are on the order of 100:1.

Other factors controlling rates and specificity of hybridization are salt and formamide concentrations and temperature, in both the hybridization and subsequent wash steps [141]. The melting or denaturation temperature (T^m) of DNA is 90–100°C. Such high temperatures applied to intact cells or chromosomes destroy their morphology and integrity. Formamide is used to lower the melting temperature of the DNA so that it does not unduly damage the target cells. Typically, denaturation is in 50% formamide in 2 × SSC at about 70°C. Hybridization is done in a mixture with precise concentrations of formamide, salt, buffer and probe in a humid environment at 37°C, for 4–24 h. Hybridization between closely related sequences with as little as 70% homology can occur. Therefore, precise conditions for washing away of excess probe, referred to as "stringency" of wash, are also crucial.

2.8.1 Types of Probes

FISH has become the technique of choice to detect chromosome abnormalities that are either too complex to be interpreted by banding or are below the resolution of standard chromosome banding techniques. Several types of probes, commonly in use, include:

1. *Satellite Probes*: These are probes that are homologous to repeated sequences around the centromeres of all chromosomes, the h regions of 1, 9, 16 and the Y and the satellites and short arms of acrocentric chromosomes. Alpha satellite probes for sequences that are specific to the centromeric regions of individual human chromosomes are commercially available. Exceptions are: chromosomes 1, 5 and 19; 13 and 21; 14 and 22. These three groups have probe sequences that cross-hybridize within each group and hence have been discontinued by at least one major probe distributor.
2. *Painting Probes*: These are libraries of probes that are specific for unique sequences isolated throughout the entire chromosome. Such libraries usually have non-specific repeated sequences repressed or removed. Paints specific for each human chromosome are commercially available.
3. *Locus-specific Probes*: Microdeletions that involve loss of segments of chromosomes less than a few megabases are usually not detectable by banding but are detectable by FISH when the appropriate probe is available. Several criteria should be met for such a probe to be useful: (1) it must be specific for a gene region associated with disease; (2) it must have been tested on enough cases to confirm specificity (frequency of association with the disease in question) and sensitivity (frequency of false positive and/or false negative results) [143]. Commercially available probes exist for about a dozen microdeletion syndromes and for an increasing number of chromosome regions involving oncogenes in cancer or leukemia.

4. *Subtelomeric Probes*: These are probes for sequences 70–300 kb in length that are immediately adjacent to the telomeres themselves and are specific for each chromosome arm (except the short arms of the acrocentric chromosomes) [144]. The forty-one different probes that are available commercially are typically used as a panel to rule out subtle structural deletions or rearrangements involving the ends of the chromosome arms.
5. *Telomeric Probe Sequence*: One specific repeated sequence, (TTAGGG)n is present at the end of every chromosome arm. The number of repeats (n) for each arm varies greatly from a minimal modal number of 300 bp to 17 kb, depending on tissue, differentiation, age and genetic factors. A critical number of repeats on each arm are necessary for the chromosome to be stable and for DNA replication of both strands to be completed without gradual loss of DNA over time [145].

2.8.2 Applications

A wide variety of applications of *in situ* hybridization techniques have been developed. The principles as outlined above are the same for all of them. Only the combinations of probes and their targets are changed. With increasing complexity of the technology, detection and data analysis is typically augmented by special computer software.

1. *Dual-colored Probes*: In the case of microdeletions or other locus-specific probes, a control probe of a different color is typically included in the probe mixture and hybridized at the same time. Detection can either be with a triple band-pass filter that allows detection of three different wavelengths (three different colors) simultaneously, or with three different single band-pass filters that each allows detection of only a single color at a time. In the latter case, individually collected digitized images are typically superimposed by computer software to generate a single three-colored image [146, 147].
2. *Multiple-colored Probes*: Several available systems allow visualization of the entire genome in multiple colors, accomplished by labeling DNA representing a particular chromosome in three or more colors and combining these colors in different ratios to give a different color for each chromosome [148]. Such combinatorial labeling can be achieved by superimposing narrow band-pass filter images that allow distinction of the various color ratios (so-called M FISH) [149] or by quantitatively measuring the pixel-value of each color and assigning a new color for each of the ratios (so-called SKY-FISH) [150, 151]. A third more esoteric method combined color ratio labeling of individual orangutan chromosomes and inter-species hybridization of the multicolored orangutan DNA to human metaphases with a resultant multi-colored banding pattern on the human chromosomes representing rearrangement of the orangutan genome in its evolution to the human karyotype (so-called Rx-FISH) [152].
3. *Comparative Genomic Hybridization*: This method of hybridization has been used mainly to characterize complex multiple chromosome abnormalities in

tumor cell lines [153–155]. DNA from the tumor line is extracted and labeled with a green fluorescent tag such as FITC and is mixed in equal molar amounts with DNA extracted from a normal cell line that has been labeled with a red fluorescent tag such as Texas red or rhodamine. The probe mixture is then hybridized to metaphases prepared from normal cells. Segments of chromosomes or entire chromosomes that are either in excess or deleted will have more or fewer green- vs. red-labeled sequences competing for complementary sites on the normal chromosomes. A segment that is in excess will have more green than red sequence (3:2 ratio) and produce a signal on the target chromosome that is correspondingly more green; conversely, a segment that is deleted will have less green than red sequence (1:2 ratio) and produce a signal that is correspondingly more red. Normal diploid segments have equal numbers (1:1 ratio) of red- and green-labeled sequences in the probe mixture and hence produce a yellow signal over chromosome regions that are not lost or gained. Such differences in ratios may not be seen easily in a single metaphase by eye and so are typically measured spectrophotometrically and the results combined from multiple metaphases.

2.8.3 Studies of Heteromorphisms by FISH

This section deals directly with the molecular characterization, particularly of the satellite DNA's that make up the most visible, structurally variable regions of the human genome. Such studies, for the most part, are based on results from a few cases and do not attempt to correlate molecular and cytological findings in any significant population of normal individuals. With one or two exceptions [156–158], the characterization of heteromorphisms at the molecular level is more anecdotal than systematic. However, more accurate characterization of heteromorphisms detected by banding is greatly augmented by the application of molecular cytogenetic techniques. FISH allows identification of specific segments of DNA in ways that are not possible with any of the standard ways of studying chromosome by conventional banding techniques. At the same time, they can detect new forms of heteromorphism in the human genome that were not detectable by previous methods. In fact, one of the drawbacks of FISH technologies is that variation in signal size or the apparent lack of a signal with a probe that is associated with a certain disease could be reflecting normal variability instead. Therefore, care must be taken in the interpretation of results when a new probe is used or the disease has not been well characterized [143]. It is also important to realize that differences in signal size by FISH, is more qualitative than quantitative. The smaller the signal, the more it may vary in size and frequency of detection.

Regions in the human genome that are variable in size and staining properties are also heterogeneous in their repetitive DNA make-up. Secondary constrictions of chromosomes 1, 9 and 16 are good examples of such heterogeneity, where reshuffling of repetitive sequences frequently occurs and can be detected by FISH. Because such sequences are closely related, there also can be frequent cross-hybridization.

References

1. Denver Conference (1960/1966) A proposed standard system of nomenclature of human mitotic chromosomes. Lancet 1:1063–1065; reprinted in Chicago Conference, pp 12–15
2. London Conference (1963) Normal human karyotype. Cytogenetics 2:264–268; reprinted in Chicago Conference 1966, pp 18–19
3. Chicago Conference (1966) Standardization in human cytogenetics. Birth defects: original article series, vol 2, no 2. The National Foundation, New York
4. Caspersson T, Farber S, Foley GE, Kudynowski J, Modest EJ, Simonsson E, Wagh U, Zech L (1968) Chemical differentiation along metaphase chromosomes. Exp Cell Res 49:219–222
5. Caspersson T, Zech L, Johansson C (1970) Analysis of human metaphase chromosome set by aid of DNA-binding fluorescent agents. Exp Cell Res 62:490–492
6. Geraedts JPM, Pearson PL (1974) Fluorescent chromosome polymorphisms; frequencies and segregation in a Dutch population. Clin Genet 6:247–257
7. Lin CC, Gideon MM, Griffith P, Smink WK, Newton DR, Wilkie L, Sewell LM (1976) Chromosome analysis on 930 consecutive newborn children using quinacrine fluorescent banding technique. Hum Genet 31:315–328
8. Lubs HA, Patil SR, Kimberling WJ, Brown J, Cohen M, Gerald P, Hecht F, Myrianthopoulos N, Summit RL (1977) Q and C-banding polymorphisms in 7 and 8 year old children: racial differences and clinical significance. In: Hook E, Porter IH (eds) Population cytogenetic studies in humans. Academic Press, New York, pp 133–159
9. Craig-Holms AP, Moore FB, Shaw MW (1973) Polymorphism of human C-band heterochromatin I: frequency of variants. Am J Hum Genet 25:181–192
10. Müller HJ, Klinger HP, Glasner M (1975) Chromosome polymorphism in a human newborn population II. Potentials of polymorphic chromosome variants for characterizing the idiogram of an individual. Cytogenet Cell Genet 15:239–235
11. McKenzie WH, Lubs HA (1975) Human Q and C chromosomal variations: distribution and incidence. Cytogenet Cell Genet 14:97–115
12. Magenis RE, Palmer CG, Wang L, Brown M, Chamberlin J, Parks M, Merritt AD, Rivas M, PL Yu (1977) Heritability of chromosome banding variants. In: Hook EB, Porter IH (eds) Population cytogenetics. Studies in humans. Academic Press, New York, pp 179–188
13. Dutrillaux B, Lejeune J (1971) Cytogénétique humaine. Sur une nouvelle technique d' analyse du caryotype humain. CR Acad Sci 272:2638–2640
14. Howell WM, Denton TE, Diamond JR (1975) Differential staining of the satellite regions of human acrocentric chromosomes. Experentia 31:260–262
15. Bobrow M, Madan K, Pearson PL (1972) Staining of some specific regions of human chromosomes, particularly the secondary constriction of no. 9. Nat New Biol 238:122–124
16. Paris Conference (1971/1972) Standardization in human genetics. Birth defects original article series, vol 8, no 7. The National Foundation, New York
17. Paris Conference Supplement (1975) Standardization in human genetics. Birth defects original article series, vol XI, no 9. The National Foundation, New York
18. ISCN (2009) In: Shafer LG, Tommerup N (eds) An international system for human cytogenetic nomenclature. S Karger, Basel
19. Saunders GF, Hsu TC, Getz MJ, Simes EL, Arrighi F (1972) Locations of human satellite DNA in human chromosomes. Nat New Biol 236:244–246
20. Jones KW, Corneo G (1971) Location of satellite and homogeneous DNA sequences on human chromosomes. Nat New Biol 233:268–271
21. Ginelli E, Corneo G (1976) The organization of repeated DNA sequences in the human genome. Chromosoma (Berl) 56:55–69
22. Verma RS, Babu A (1995) Human chromosomes. Principles and techniques. McGraw-Hill, New York, pp 72–127
23. Babu A, Macera MJ, Verma RS (1986) Intensity heteromorphisms of human chromosome 15p by DA/DAPI technique. Hum Genet 73:298–300

24. Corneo G, Ginelli E, Polli E J (1968) Isolation of complementary strands of a human satellite DNA. J Mol Biol 33:331
25. Miklos GLG, John B (1979) Heterochromatin and satellite DNA in man: properties and prospects. Am J Hum Genet 31:264–280
26. Gosden JR, Mitchell AR, Buckland RA, Clayton RP, Evans HJ (1975) The location of four human satellite DNAs on human chromosomes. Exp Cell Res 92:148–158
27. Jones KW, Prosser J, Corneo G, Ginelli E (1973) The chromosomal localization of human satellite DNA III. Chromosoma (Berl) 42:445–451
28. Buhler EM, Tsuchimoto T, Jurik LP, Stalder GR (1975) Satellite DNA III and alkaline Giemsa staining. Humangenetik 26:329–333
29. Levi B, Warburton P (2004) Molecular dissection of heteromorphic regions. In: Wyandt HE, Tonk VS (eds) Atlas of human chromosome heteromorphisms. Kluwer, Dordrecht, pp 97–105
30. Hopman AHN, Raap AK, Landegent JE, Wiegant J, Boerman RH, Van Der Ploeg M(1988) Non radioactive *in situ* hybridization. In: Van Leeuwen FW et al (eds) Molecular neuroanatomy. Elsevier, Amsterdam, pp 43–68
31. Lichter P, Ried T (1994) Molecular analysis of chromosome aberrations. *In situ* hybridization. In: Gosden JR (ed) Methods in molecular biology. Chromosome analysis protocols. Humana Press, Totowa, pp 449–478
32. Caspersson T, Zech L, Johansson C (1970) Analysis of human metaphase chromosome set by aid of DNA-binding fluorescent agents. Exp Cell Res 62:490–492
33. Weisblum B, de Haseth PL (1972) Quinacrine, a chromosome stain specific for deoxyadenylate-deoxythymidylate rich regions in DNA. Proc Natl Acad Sci U S A 63:629–632
34. Ellison JR, Barr HJ (1972) Quinacrine fluorescence of specific chromosome regions. Late replication and high A:T content in Samoia leonensis. Chromosoma 36:375–390
35. Comings DE, Kovacs BW, Avelino E, Harris DC (1975) Mechanism of chromosome banding. V. Quinacrine banding. Chromosoma 50:111–145
36. Michelson AM, Monny C, Kovoor A (1972) Action of quinacrine mustard on polynucleotide. Biochimie 54:1129–1136
37. Sumner AT, Evans HJ, Buckland RA (1971) New technique for distinguishing between human chromosomes. Nat New Biol 232:31–32
38. Drets ME, Shaw MW (1971) Specific banding patterns of human chromosomes. Proc Nat Acad Sci U S A 68:2073–2077
39. Seabright M (1971) A rapid banding technique for human chromosomes. Lancet 2:971–972
40. Wang HC, Fedoroff S (1972) Banding in human chromosomes treated with trypsin. Nat New Biol 235:52–54
41. Müller W, Rozenkranz W (1972) Rapid banding technique for human and mammalian chromosomes. Lancet 1:898
42. Pearson P (1972) The use of new staining techniques for human chromosome identification. J Med Genet 9:264–275
43. Sumner AT (1994) Chromosome banding and identification absorption staining. In: Gosden GR (ed) Methods in molecular biology. chromosome analysis protocols. Humana Press, Totowa, pp 59–81
44. Saitoh Y, Laemmli K (1994) Metaphase chromosome structure: bands arise from a differential folding path of highly AT-rich scaffold. Cell 75:609–661
45. Hancock R (2000) A new look at the nuclear matrix. Chromosoma 109(4):219–225
46. Nickerson J (2001) Experimental observations of the nuclear matrix. J Cell Sci 114(3):463–474
47. Comings DE, Avelino AX (1974) Mechanisms of chromosome banding II. Evidence that histones are not involved. Exp Cell Res 86:202–206
48. Comings DE, Avelino E, Okada TA, Wyandt HE (1973) The mechanisms of C- and G-banding of chromosomes. Exp Cell Res 77:469–493
49. Sumner AT (1973) Involvement of protein disulphides and sulphydryls in chromosome banding. Exp Cell Res 83:438–442
50. Korf BR, Schuh BE, Salwen MJ, Warburton D, Miller OJ (1976) The role of trypsin in the pre-treatment of chromosomes for Giemsa banding. Hum Genet 31:27–33

51. Holmquist GP (1992) Chromosome bands, their chromatin flavors and their functional features. Am J Hum Genet 51:17–37
52. Brown RL, Pathak S, Hsu TC (1975) The possible role of histones in the mechanism of G-banding. Science 189:1090–1091
53. Holmquist GP, Comings DE (1976) Histones and G-banding of chromosomes. Science 193:599–602
54. Retief AE, Ruchel R (1977) Histones removed by fixation. Their role in the mechanism of chromosome banding. Exp Cell Res 106:233–237
55. Okada TA, Comings DE (1974) Mechanisms of chromosome banding III. Similarity between G-bands of mitotic chromosomes and chromomeres of meiotic chromosomes. Chromosoma 48:65–71
56. Jhanwar SC, Burns JP, Alonso ML, Hew W, Chaganti RS (1982) Mid-pachytene chromomere maps of human autosomes. Cytogenet Cell Genet 33:240–248
57. Comings DE (1975) Mechanisms of chromosome banding. IV. Optical properties of the Giemsa dyes. Chromosoma 50:89–110
58. Comings DE, Avelino E (1975) Mechanisms of chromosome banding. VII. Interaction of methylene blue with DNA and chromatin. Chromosoma 51:365–379
59. Wyandt HE, Anderson RS, Patil SR, Hecht F (1980) Mechanisms of Giemsa banding. II. Giemsa components and other variables in G-banding. Hum Genet 53:211–215
60. Dutrillaux B, Lejeune J (1971) Cytogenetique humaine. Sur une nouvelle technique d' analyse du caryotype humain. CR Acad Sci (III) 272:2638–2640
61. Bobrow M, Collacott HEAC, Madan K (1972) Chromosome banding with acridine orange. Lancet 2:1311
62. Wyandt HE, Vlietinck RF, Magenis RE, Hecht F (1974) Colored reverse-banding of human chromosomes with acridine orange following alkaline-formalin treatment: densitometric validation and applications. Humangenetik 23:119–130
63. Verma RS, Babu A (1995) Human chromosomes. Principles and techniques, 2nd edn. McGraw-Hill, New York
64. Pardue ML, Gall JG (1970) Chromosomal localization of mouse satellite DNA. Science 168:1356–1358
65. Arrighi FE, Hsu TC (1971) Localization of heterochromatin in human chromosomes. Cytogenetics 10:81–86
66. Hsu TC, Arrighi FE (1971) Distribution of constitutive heterochromatin in mammalian chromosomes. Chromosoma 34:243–253
67. Sumner AT (1972) A simple technique for demonstrating centromeric heterochromatin. Exp Cell Res. 75:304–306
68. McKenzie WH, Lubs HA (1973) An analysis of the technical variables in the production of C bands. Chromosoma 41:175–182
69. Eiberg H (1974) New selective Giemsa technique for human chromosomes, Cd staining. Nature 248:55
70. Nakagome Y, Abe T, Misawa S, Takeshita T, Linuma K (1984) The "loss" of centromeres from chromosomes of aged women. Am J Hum Genet 36:398–404
71. Nakagome Y, Nakahori Y, Mitani K, Matsumoto M (1986) The loss of centromeric heterochromatin from an inactivated centromere of a dicentric chromosome. Jinrui Idengaku Zasshi 31:21–26
72. Evans HJ, Ross A (1974) Letter: spotted centromeres in human chromosomes. Nature 249:861–862
73. Maraschio P, Zuffardi O, Lo CF (1980) Cd bands and centromeric function in dicentric chromosomes. Hum Genet 54:265–267
74. Willard HF (1990) Centromeres of mammalian chromosomes. Trends Genet 6:410–416
75. Patil SR, Merrick S, Lubs HA (1971) Identification of each human chromosome with a modified Giemsa stain. Science 173:821–822
76. Bobrow M, Madan K, Pearson PL (1972) Staining of some specific regions of human chromosomes, particularly the secondary constriction of no. 9. Nat New Biol 238:122–124

77. Wyandt HE, Wysham DG, Minden SK, Anderson RS, Hecht F (1976) Mechanism of Giemsa banding of chromosomes I. Giemsa-11 banding with azure and eosin. Exp Cell Res 102:85–94
78. Howell WM, Denton TE, Diamond JR (1975) Differential staining of the satellite regions of human acrocentric chromosomes. Experentia 31:260–262
79. Howell WM, Black DA (1980) Controlled silver-staining of nucleolus organizer regions with a protective colloidal developer: A 1-step method. Experentia 36:1014–1015
80. Miller OJ, Miller DA, Dev VG, Tantravahi R, Croce CM (1976) Expression of human and suppression of mouse nucleolus organizer activity in mouse-human somatic cell hybrids. Proc Natl Acad Sci U S A 73:4531–4535
81. Goodpasture C, Bloom SE, Hsu TC, Arrighi FE (1976) Human nucleolus organizers: the satellites or the stalks? Am J Hum Genet 28:559–566
82. Verma RS, Rodriguez J, Shah JV, Dosik H (1983) Preferential association of nucleolar organizing human chromosomes as revealed by silver staining technique at mitosis. Mol Gen Genet 190:352–354
83. Perry P, Wolff S (1974) New Giemsa method for differential staining of sister chromatids. Nature 261:156–158
84. Latt SA (1973) Microfluorometric detection of deoxyribonucleic acid replication in human metaphase chromosomes. Proc Natl Acad Sci U S A 70:3395–3399
85. Gebhart E (1981) Sister chromatid exchange (SCE) and structural chromosome aberration in mutagenicity testing. Hum Genet 58:235–254
86. Craig-Holmes AP, Shaw MW (1976) Cell cycle analysis of asynchronous cultures using the BudR-Hoechst technique. Exp Cell Res 99:79–87
87. Crossen PE, Morgan WF (1977) Analysis of human lymphocyte cell cycle time in culture measured by sister chromatid differential staining. Exp Cell Res 104:453–457
88. German J, Crippa LP, Bloom D (1974) Bloom's syndrome. III. Analysis of the chromosome aberrations characteristic of this disorder. Chromosoma 48:361–366
89. Lyon MF (1961) Gene action in the X-chromosome of the mouse (Mus musculus L.). Nature 190:372–373
90. German JL III (1962) DNA synthesis in human chromosomes. Trans N Y Acad Sci 24:395–407
91. Peterson AJ (1964) DNA synthesis and chromosomal asynchrony. J Cell Biol 23:651–654
92. Priest JH, Heady JE, Priest RE (1967) Delayed onset of replication of human X chromosomes. J Cell Biol 35:483–486
93. Lyon MF (1972) X-chromosome inactivation and developmental patterns in mammals. Biol Rev 47:1–35
94. Latt SA (1975) Fluorescence analysis of late DNA replication in human metaphase chromosomes. Somat Cell Genet 1:293–321
95. Angell RR, Jacobs PA (1975) Lateral asymmetry in human constitutive heterochromatin. Chromosoma 51:301–310
96. Lin MS, Alfi OS (1978) Detection of lateral asymmetry in the C band of human chromosomes by BrdU-DAPI fluorescence. Somatic Cell Genet 4:603–608
97. Angell RR, Jacobs PA (1978) Lateral asymmetry in human constitutive heterochromatin: frequency and inheritance. Am J Hum Genet 30:144–152
98. Gosh PK, Rani R, Nand R (1979) Lateral asymmetry of constitutive heterochromatin in human chromosomes. Hum Genet 52:79–84
99. Yunis JJ (1976) High resolution of human chromosomes. Science 191:1268–1269
100. Yunis JJ (1981) Mid prophase human chromosomes; the attainment of 2000 bands. Hum Genet 56:295–298
101. Ikeuchi T (1984) Inhibitory effect of ethidium bromide on mitotic chromosome condensation and its application to high-resolution chromosome banding. Cytogenet Cell Genet 38:56–61
102. Balicek P, Zizka J (1980) Intercalar satellites of human acrocentric chromosomes as a cytological manifestation of polymorphisms in GC-rich material. Hum Genet 54:343–347

103. Levi B, Warburton P (2004) Molecular dissection of heteromorphic regions. In: Wyandt HE, Tonk VS (eds) Atlas of human chtomosome heteromorphisms. Kluwer, Dordrecht
104. Waye JS, Creeper LA, Willard HF (1987) Organization and evolution of alpha satellite DNA from human chromosome 11. Chromosoma 95:182–188
105. Choo KHA, Vissel B, Earle E (1989) Evolution of Alpha satellite DNA on human acrocentric chromosomes. Genomics 5:332–344
106. Waye JS, Willard HF (1989) Human beta satellite DNA: genomic organization and sequence definition of a class of highly repetitive tandem DNA. Proc Natl Acad Sci U S A 86:6250–6254
107. Vissel B, Choo KH (1989) Mouse major (gamma) satellite DNA is highly conserved and organized into extremely long tandem arrays: implications for recombination between non-homologous chromosomes. Genomics 5:407–414
108. Wier HU, Zitzelsberger HF, Gray JW (1992) Differential staining of human and murine chromatin *in situ* by hybridization with species-specific satellite DNA probes. Biochem Biophys Res Commun 182:1313–1319
109. Lee C, Li X, Jabs EW, Court D, Lin CC (1995) Human gamma X satellite DNA: an X chromosome specific centromeric DNA sequences. Chromosoma 104:103–112
110. Johnson DH, Kroisel PM, Klapper HJ, Rosenkranz W (1992) Microdissection of a human marker chromosome reveals its origin and a new family of centromeric repetitive DNA. Hum Mol Genet 1:741–747
111. Corneo G, Ginelli E, Polli E (1970) Repeated sequences in human DNA. J Mol Biol 48:319–327
112. Jones KW, Prosser J, Corneo G, Ginnelli E, Bobrow M (1973) Satellite DNA, constitutive heterochromatin and human evolution. In: Pfeiffer RA (ed) Modern aspects of cytogenetics: constitutive heterochromatin in man. Schattauer, Stuttgart, pp 45–61
113. Jones KW, Purdom LF, Prosser J, Corneo G (1974) The chromosomal location of human satellite DNA I. Chromosoma 49:161–171
114. Prosser J, Frommer M, Paul C, Vincent PC (1986) Sequence relationships of three human satellite DNAs. J Mol Biol 187:145–155
115. Tagarro I, Wiegant J, Raap AK, Gonzalez-Aguilera JJ, Fernandez-Peralta AM (1994) Assignment of human satellite 1 DNA as revealed by fluorescent *in situ* hybridization with oligonucleotides. Hum Genet 93:125–128
116. Jeanpierre M (1994) Human satellites 2 and 3. Ann Genet 37:63–71
117. Jackson MS, Mole SE, Ponder BA (1992) Characterization of a boundary between satellite III and alphoid sequences on human chromosome 10. Nucleic Acids Res 20:4781–4787
118. Willard HF (1991) Evolution of alpha satellite DNA. Curr Opin Genet Dev 1:509–514
119. Sullivan BA, Schwartz S (1995) Identification of centromeric antigens in dicentric Robertsonian translocations: CENP-C and CENP-E are necessary components of functional centromeres. Hum Mol Genet 4:2189–2197
120. Gieni RS, Chan GKT, Hendzel MJ (2008) Epigenetics regulate centromere formation and kinetochore function. J Cell Biochem 104:2027–2039
121. Silva MCC, Jansen LET (2009) At the right place at the right time: novel CENP-A binding proteins shed light on centromere assembly. Chromosoma 118:567–574
122. Bosi PR, Grant GR, Jeffreys AJ (2002) Minisatellites show rare and simple intra-allelic instability in the mouse germ line. Genomics 80:2–4
123. Desmarais E, Vigneron S, Buresi C, Cambien F, Cambou JP, Roizes G (1993) Variant mapping of the Apo(B) AT rich minisatellite. Dependence on nucleotide sequence of the copy number variations. Instability of the non-canonical alleles. Nucleic Acids Res 21:2179–2184
124. Buresi C, Desmarais E, Vigneron S, Lamarti H, Smaoui N, Cambien F, Roizes G (1996) Structural analysis of the minisatellite present at the 3′ end of the human apolipoprotein B gene: new definition of the alleles and evolutionary implications. Hum Mol Genet 5:61–68
125. Yu S, Mangelsdorf M, Hewett D, Hobson L, Baker E, Eyre HJ, Lapsys N, Le Paslier D, Doggett NA, Sutherland GR, Richards RI (1997) Human chromosomal fragile site FRA16B is an amplified AT-rich minisatellite repeat. Cell 88:367–374

126. Hewett DR, Handt O, Hobson L, Mangelsdorf M, Eyre HJ, Baker E, Sutherland GR, Schuffenhauer S, Mao JI, Richards RI (1998) FRA10B structure reveals common elements in repeat expansion and chromosomal fragile site genesis. Mol Cell 1:773–781
127. Jobling MA, Bouzekri N, Taylor PG (1998) Hypervariable digital DNA codes for human paternal lineages: MVR-PCR at the Y-specific minisatellite, MSY1(DYF155S1). Hum Mol Genet 7:643–653
128. Butler JM (2004) Short tandem repeat analysis for human identity testing. Current protocols in human genetics, Unit 14.8 (Supplement 41). Wiley, Hoboken, pp 14.8.1–14.8.22
129. Butler JM (2007) Short tandem repeat typing technologies used in human identity testing. Biotechniques43(Suppl):i–iv
130. Payseur BA, Jing P (2009) A genomewide comparison of population structure at STRPs and nearby SNPs in humans. Mol Biol Evol 26:1369–1377
131. Manolio TA, Collins FS (2009) The HapMap and genome-wide association studies in diagnosis and therapy. Ann Rev Med 60:443–456
132. Efstratiadis A, Kafatos FC, Mazam AM, Maniatis T (1976) Enzymztic in vitro synthesis of globin genes. Cell 7:279–288
133. White TJ, Arnheim N, Erlich HA (1989) The polymerase chain reaction. Trends Genet 5:185–189
134. Sambrook J et al (1989) Molecular cloning: a laboratory manual. Cold Spring Harbor Laboratory, New York
135. Shero JH, McCormick MK, Antonarakis SE, Hieter P (1991) Yeast artificial chromosome vectors for efficient clone manipulation and mapping. Genomics 10:505–508
136. Monaco AP, Larin Z (1994) YACs, BACs and MACs: artificial chromosomes as research tools. Trends Biotechnol 12:280–286
137. Ludecke HJ, Senger G, Claussen U, Horsthemke B (1989) Cloning defined regions of the human genome by microdissection of banded chromosomes and enzymatic amplification. Nature 338:348–350
138. Kao F-T, Yu J-W (1991) Chromosome microdissection and cloning in human genome and genetic disease analysis. Proc Natl Acad Sci U S A 88:1844–1848
139. Harris P, Boyd E, Ferguson-Smith MA (1985) Optimizing human chromosome separation for the production chromosome-specific DNA libraries by flow sorting. Hum Genet 70:59–65
140. Deaven LL, van Dilla MA, Bartholdi MF, Carrano AV, Cram LS, Fuscoe JC, Gray JW, Hildebrand CE, Moyzis RK, Perlman J (1986) Construction of human chromosome-specific DNA libraries from flow-sorted chromosomes. Cold Spring Harb Symp Quant Biol 51:159–167
141. Hopman AHN, Raap AK, Landegent JE, Wiegant J, Boerman RH, Van der Ploeg M (1988) Non radioactive *in situ* hybridization. In: Van Leeuwen FW et al (eds) Molecular neuroanatomy. Elsevier, Amsterdam, pp 43–68
142. Lichter P, Ried T (1994) Molecular analysis of chromosome aberrations. *In situ* hybridization. In: JR Gosden (ed) Methods in molecular biology. Chromosome analysis protocols. Humana Press, Totowa, pp 449–478
143. ACMG (1999) Standards and guidelines for clinical genetics laboratories, 2nd edn. American College of Medical Genetics, Bethesda
144. Knight SJ, Flint J (2000) Perfect endings: a review of subtelomeric probes and their use in clinical diagnosis. J Med Genet 57:401–409
145. Moyzis RK, Buckingham JM, Cram LS, Dani M, Deaven LL, Jones MD, Meyne J, Ratliff RL, Wu J-L (1988) A highly conserved repetitive sequence (TTAGGG)n, present at the telomeres of human chromosomes. Proc Natl Acad Sci U S A 85:6622–6626
146. Nederlof PM, Robinson D, Abuknesh R, Weigant J, Hopman AHN, Tanke HJ, Raap AK (1989) Three-color fluorescence *in situ* hybridization for the simultaneous detection of multiple nucleic acid sequences. Cytometry 10:20–27

147. Lichter P, Tang CJ, Call K, Hermanson G, Evans GA, Housman D, Ward DC (1990) High resolution mapping of human chromosome 11 by *in situ* hybridization with cosmid clones. Science 247:64–69
148. Dauwerse JG, Wiegant J, Raap AK, Breuning MH, van Ommen GJ (1992) Multiple colors by fluorescence *in situ* hybridization using ratio-labelled DNA probes create a molecular karyotype. Hum Mol Genet 1:593–598
149. Spiecher MR, Ballard S, Ward, DC (1996) Karyotyping human chromosomes by combinatorial multifluor FISH. Nat Genet 12:368–375
150. Schrock E, Veldman T, Padilla-Nash H, Ning Y, Spurbeck J, Jalal S, Shaffer LG, Papenhausen P, Kozma C, Phelan MC, Kjeldsen E, Schonberg SA, Obrien P, Biesecker L, du Manoir S, Reid T (1997) Spectral karyotyping refines cytogenetic diagnsotic of constitutional chromosomal abnormalities. Hum Genet 101:255–262
151. Bayani J, Squire JA (2001) Advances in the detection of chromosomal aberrations using spectral karyotyping. Clin Genet 59:65–73
152. Müller S, O'Brien PC, Feguson-Smith MA, Weinberg J (1998) Cross-species colour segmenting: a novel tool in human karyotype analysis. Cytometry 33:445–452
153. Kallioniemi A, Kallioniemi OP, Sudar D, Rutovitz D, Gray JW, Waldman F, Pinkel D (1992) Comparative genomic hybridization for molecular cytogenetic analysis of solid tumors. Science 258:818–821
154. Kallioniemi OP, Kallioniemi A, Piper J, Isola J, Waldman F, Gray J, Pinkel D (1994) Optimizing comparative genomic hybridization for analysis of DNA sequence copy number changes in solid tumors. Genes Chromosome Canc 10:231–234
155. Levy B, Dunn TM, Kaffe S, Kardon N, Hirschhorn K (1998) Clinical applications of comparative genomic hybridization. Genet Med 1:4–12
156. Bossuyt PJ, Van Tienen MN, De Gruyter L, Smets V, Dumon J, Wauters JG (1995) Incidence of low-fluorescence alpha satellite region on chromosome 21 escaping detection of aneuploidy at interphase by FISH. Cytogenet Cell Genet 68:203–206
157. Stergianou K, Gould CP, Waters JJ, Hulten MA (1993) A DA/DAPI positive human 14p heteromorphism defined by fluorescent *in situ* hybridization using chromosome 15-specific probes D15Z1 (Satellite III) and p-TRA-25 (alphoid). Hereditas 119:105–110
158. Verlinsky Y, Ginsberg N, Chmura M, Freidine M, White M, Strom C, Kuliev A (1995) Cross-hybridization of the chromosome 13/21 alpha satellite DNA probe to chromosome 22 in the prenatal screening of common chromosomal aneuploidies by FISH. Prenat Diagn 15:831–834

Chapter 3
Frequencies of Heteromorphisms

An initial attempt to assess the frequency of variants was made in non-banded chromosomes from consecutive newborns by Lubs and Ruddle [1]. Their study included 3,476 infants of white mothers and 807 infants of black mothers, all of whom were phenotypically normal except one child with low birth weight. A total of 2,131 variants involving chromosomes A1, C9, E16, the short arms and satellites of D and G group chromosomes, and Y long arm were scored. Frequencies of certain striking variants were found to be discrepant between black and white children (Table 3.1). In particular, a metacentric C9 variant (later recognized as a 9qh inversion) was 20 times more frequent in the black children; a large short arm on a D-group chromosome was four times more frequent. Y chromosome length was not different for black and white children. However, a large Y (>E18) was present in one of nine Chinese infants included in the study and a second large Y was present in the only Turkish infant. Earlier studies had shown a high frequency of large Y in Japanese adult males [2].

3.1 By Q- and C-banding

With the development of chromosome banding techniques in the early 1970s, numerous studies were done to determine the frequencies of variants in the general population by banding [3–11]. Frequencies of minor variants by Q and/or C-banding in various populations, not surprisingly, showed differences due to ethnic origin, age distribution and ascertainment. However, criteria used in these studies were often subjective and the populations selected often introduced biases, making it difficult to directly compare frequencies. Nevertheless, there emerged recognition of the chromosome regions that are the most variable.

One of the first population assessments of Q-band variants was by Geraedts and Pearson [7] in 221 Dutch individuals: 132 individuals were from 16 families; 75 individuals were from additional 3 families. Q-banding had the advantage that every chromosome could be identified. Furthermore, the variants allowed distinction between many pairs of homologs. Variants were first scored in the oldest generation of

H. E. Wyandt, V. S. Tonk, *Human Chromosome Variation: Heteromorphism and Polymorphism*, DOI 10.1007/978-94-007-0896-9_3,

Table 3.1 Frequencies (%) of variants in Caucasian and Black infants in unbanded chromosomes. (Reprinted by permission from Macmillan Publishers Ltd: Lubs and Ruddle (1971). Nature 233: 134–136 [2])

Description of variant	Caucasian N=3,476	Black N=807
1qh+	0.23	0.37
metacentric (C9)	0.06	1.24
Dp+ (=E18p)	13.80	21.18
Dp+ (>E18p)	0.25	0.87
Ds+ (>Dp)	2.24	4.21
Dss (tandem s)	0.09	0.25
Dp−	0.06	0
E16+ (>C6p)	2.04	3.96
E16<E18	0.55	0.37
Gp+ (=E18p)	2.33	5.58
Gp+ (>E18p)	0.06	0
Gs+ (>Gp)	2.65	4.46
Gss (tandem s)	0.09	0
Totals=	24.45	42.49
Y (>F19)	14.53	14.59
Y (>E18)	0.34	0.24
Y (<G22)	0.40	0.48
Y metacentric	0.11	0

the 19 pedigrees with variants passed on from the mother coded as 1 and those from the father as 2. New polymorphisms, introduced in the family by marriage, were also coded. Chromosomes lacking of a polymorphism were left non-coded. The frequency of Q-band variants by this system was approximately 4 per individual with no significant difference between males and females. Excluding chromosomes 1, 9, 16 and the Y, the highest numbers of heteromorphisms were in chromosomes 3, and 13. The numbers of heterozygotes vs. homozygotes were also scored for each chromosome. Hardy-Weinberg expectations were calculated and were met for chromosomes 3, 4, 14, 21 and 22. However, for chromosomes 13 and 15 there was an excess of heterozygotes and a lack of homozygotes. The excess of heterozygotes of chromosome 13 appeared to come from the maternal side in the case of sons and from the paternal side in the case of daughters. Of interest in this regard, Fogle and McKenzie [12] studied 81 members of a black kindred by sequential Q- and C-banding. Preferential segregation of Q-heteromorphisms was found and was especially distorted for chromosome 13.

Mackenzie and Lubs [13] studied variants in 77 normal newborns from Grand Junction, Colorado. C-banding was preceded by Q- or G-banding, in some cases on the same cells so that variants on homologs by C-banding could be distinguished. Q- and C-band variants were classified by intensity and size, respectively. Of 391 variants described, 57.6% were Q-band and 42.4% were C-band variants. Q-band variants, excluding the Y, were restricted to autosomes 3, 4, 13–15, 21 and 22. C-band variants, while fewer in number were found to give at least one variant in every chromosome. Except for giant satellites on the acrocentric chromosomes,

Q-band and C-band variants often represented different chromosome regions. Overall, an average of approximately 5 heteromorphisms was found per individual.

Müller et al. [14] studied polymorphisms, by C-, Q- and G-banding, in 376 neonates in a New York hospital. Classification of variants was according to the 1971 Paris Conference (Table 2.1). C-band variants of 1, 9 and 16 were compared to the length of the long arm of chromosome 21. Infants' karyotypes were first examined by G-banding; those with a chromosome abnormality were excluded from the study. For those included, metaphases were examined by Q-banding alone, by C-banding alone and by Q- and C-banding of the same metaphases. The final classification of variants was made from the latter category. For chromosomes 1, 9 and 16, variations were seen in size or in the position of the centromere. The remaining chromosomes were scored for size of heterochromatin by C-banding and for intensity by Q-banding. C-banding revealed apparent partial inversions of heterochromatin into the short arms in 1.6% of no. 1s, 10.7% of no. 9s and 1.4% of no. 16s. Complete inversion of heterochromatin into the short arm occurred in about 0.6% of no. 9s. Chromosomes 6, 8, 10, 12, and the X had larger than average C-bands in <1% of subjects. Chromosomes 7 and 11 had larger bands in 10 and 13% of subjects, respectively. Chromosomes 17–20 occasionally had C-bands that were about twice the average size (0.0–2.0%). The D and G groups showed heteromorphisms in the short arm and satellite regions in every chromosome.

Lin et al. [11] studied Q-band variants in 930 consecutive newborns in a Canadian population that was 87% white and 13% black and Oriental. Variants were classified by intensity level as outlined in the Paris Conference (1971), using methods similar to those of McKenzie and Lubs. A bright variable band in chromosome 3 was present in the long arm in 77% of cases. In about 33%, this variant was present in both homologs. In<1% of cases the bright variant appeared to be inverted and was in the short arm. All cases were normal, whereas Soudek et al. [15] reported a higher incidence of this inverted bright variant in children with mental retardation (see Part IV, Chapter 32).

Mikelsaar et al. [16–18] did several detailed studies of variants by G-banding and by Q-banding in an Estonian population of normal adults and of children with mental retardation. Variants more frequent in men than women included 16q+ in adults; 17ph+, 15p+ and short arm variants of 21 in mental retardation of unknown etiology (MRU); and 15pss in Down syndrome. Variants that were more frequent in children with Down syndrome than in normal adults were 9q+ and 15pss. Variants more frequent in Down syndrome than in MRU were 13ps, 14p+, 15ps+, 15pss and 22p+. Yq++ was also more frequent in MRU, but not significantly different from normal males. One variant in a patient with Down syndrome was 14ps+ (unusually large bright satellites). Combinations of variants (multiple bright regions) in one patient and multiple G-band variants (9q+, 13p–, 17ph– and Yq–) in a second patient were unusual. Overall, Q-banding revealed no significant differences in the autosomes between sexes. Girls with MRU had a higher frequency of bright 3p11q11 than normal females or females with Downs, due to a greater number of homozygotes. Males with Down syndrome showed more frequent heterozygotes of 3p11q11 than normal men or boys with MRU. Girls with mental retardation showed a different distribution of homozygotes vs. heterozygotes for bright 4p11q11 and bright satellites on 15p than normal population. The frequencies of homozygotes for 4p11q11, 13p11q11, 13p13, 15p13 and 17p12 fit Hardy-Weinberg expectations in all populations. However, normal populations showed an excess of homozygotes for 14p13, 21p13, 22p13, 22p11 and 9q12: 14p12 was in excess in males; 21p13 and 22p13 were in excess in females; in MRU 21p13 and 22p13 were in excess in females. There were also differences between males and females for 11q11 and 16q11 in normal populations and for 15p12 in children with mental retardation. Overall, differences between males and females for all classes did not differ and an excess of variants for 13p such as reported by Geraedts and Person [7] was not found. The significance of the specific differences in distributions of variants, if any, has not been clarified in any more up-to-date studies.

3.1.1 The New Haven Study

A significant attempt to study normal variants in a standardized way was made by Lubs et al. [3] in a New Haven study of 7- and 8-year-old children who had been followed from birth. Because the study of over 4,000 newborns of white and black mothers had been initiated eight years earlier [9], before banding techniques were available, a subsample of 400 children were selected by a randomization process that included equal numbers of white and black children. Included in both groups were equal numbers with IQs ≥ 85 and <85, respectively. G-banding was done on all of the children, and assessment of variants was done by Q- and C-banding. Although the study is not especially large, it was carefully designed. Frequencies and the criteria for classifying chromosome variants, described in this study, are referred to throughout much of this book, especially in the summaries of chromosome variants in Part II.

Q and C band variants studied by Lubs et al. were classified as to intensity and/or size. Q-band intensity was divided into 5 levels according to Paris Conference criteria (Table 2.1, Fig. 3.1a). Average C-band size for chromosomes 1, 9, Y and 16 were compared to the short arm of chromosome 16 (Fig. 3.1b). Level 1 was $\leq$ one-half the length of 16p and level 5 was $>$ 2x the length of 16p. C-bands of the other chromosomes were classified as small or large. Comparisons between black and white children were made of the frequencies of heteromorphisms at the extreme levels for each chromosome (Table 3.2). Differences were found for chromosomes 4, 12, 18, 19 and 22, with the greatest difference being for chromosome 19. Ten percent of black children had a large 19cen compared to 0% of white children, although, because of the low numbers of cases, none of these differences were statistically significant. C-band variants for chromosomes 1, 9 and 16 were more frequent, but level 1 variants were still not significantly different in the two races. However, level 5 C-band variants were significantly higher (nearly twice as frequent) in the black children. Inversions in chromosomes 1 and 9 were classified into three types: 1h partial inversion, 9h partial inversion and 9h complete inversion. Only one 1qh

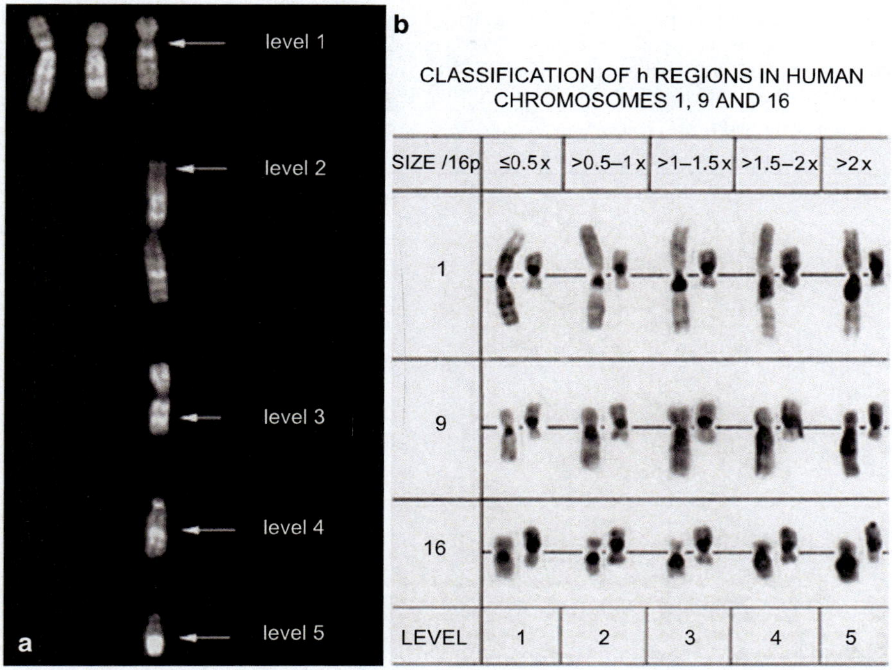

Fig. 3.1 **a** Examples of chromosome bands by QFQ-banding representing the various intensity levels (see Table 2.1) as designated by the Paris Conference, 1971 [20]. *Level 1,* represented by the primary constriction of most chromosomes (chromosomes 6, 8 and 10 are shown as examples) is the least intense. *Level 5,* represented by the distal end of the Y chromosome is the most intense. **b** Definition of the five classifications of the h regions of chromosomes 1, 9 and 16. These were coded from small to large *(1–5).* The short arm of chromosome 16 was used as the reference standard within the same cell. A code 1 was assigned to the h region, if its size was judged to less than half the size of 16p. A code 5 was assigned if it was judged to be more than twice as large as 16p. [Reproduced with permission from Elsevier: from Lubs et al. (1977). Q and C banding polymorphisms in 7 and 8 year old children: Racial differences and clinical significance. In Population Cytogenetics (Hook EB and Porter IH eds). Academic Press, NY. Fig. 2, p. 138]

partial inversion was found in Black children; 14 were found in White children. Partial inversions in 9h were more frequent in White than in black children, but the difference was not significant. However, a significantly higher frequency of complete inversion of 9qh was found in black children. No inversions in chromosome 16 were found in either race.

For Q-banding, significant differences in frequencies were seen for bright or brilliant polymorphisms (levels 4 and 5) of the centromeres for chromosomes 3, 4 and for the satellites and short arms of acrocentric chromosomes. A particularly high frequency with a large short arm on chromosome 13 was seen in the black population. Chromosomes other than 3, 4, the acrocentrics and the Y did not show polymorphisms by Q-banding. Overall, differences in frequencies of Q- and C-band polymorphisms related to IQ were not statistically significant, although large bril-

Table 3.2 Centromere polymorphism frequencies by race (C-Banding). [Reproduced with permission from Elsevier: from Lubs et al. (1977). Q and C banding polymorphisms in 7 and 8 year old children: Racial differences and clinical significance. In Population Cytogenetics (Hook EB and Porter IH eds). Academic Press, NY. Table 10, p 124]

Chromosome region	WHITE (n=95)			BLACK (N=97)		
	Small	Large	Total	Small	Large	Total
2c	1	–	1	–	3	3
3c	3	4	7	3	3	6
4c	3	–	3	3	7	10
5c	–	3	3	6	3	9
6c	2	3	5	1	–	1
7c	–	4	4	1	3	4
8c	–	1	1	–	–	–
10c	1	3	4	1	3	4
11c	6	–	6	3	1	4
12c	–	1	1	–	5	5
13c	2	3	5	5	4	9
14c	1	2	3	3	–	3
15p	–	4	4	–	5	5
15c	6	4	10	9	1	10
17c	4	2	6	2	4	6
18p	1	–	1	–	1	1
18c	9	–	9	7	8	15
19c	2	–	2	4	9	13
20c	6	1	7	1	5	6
21c	–	–	–	–	4	4
22c	1	2	3	4	8	12
Xc	–	5	5	–	1	1
Total	48	42*	90	53	78*	131

*$X^2 = 11.21$, $p < 0.005$

liant satellites on 21 were half as frequent in black children with low IQ and large brilliant satellites on 22 were twice as frequent in White children with low IQ. Again, small numbers of children were represented in each group, and no attempt was made to extend these correlations to other family members. Studies of the frequencies of heteromorphisms by others in selected populations with mental retardation, fetal wastage, aneuploidy and other clinical conditions are discussed in Sect. 4.

3.1.2 Study Comparisons

The major studies of frequencies of normal variants by Q, and C-banding are summarized in Tables 3.3 and 3.4. Although comparisons of the relative frequencies of variants involving different chromosome regions are tabulated, such comparisons

Table 3.3 Frequencies of bright to brilliant Q autosomal variants from five early studies

Chr	Geraedts and Pearson 1974 n=221[a]	Müller et al. 1975 n=376[b]	Lubs et al. 1976 n=400[c]	Lin et al. 1976 n=930[d]	Mikelsaar et al. 1976 n=349[e]
3c	48.4	54.7	46.7	55.5	55.1
4c	2.7	2.8	8.1	14.1	28.4
13p	50.0	44.2	42.9	31.3	72.9
13s		7.5	3.5	1.9	9.3
14p	14.3	2.4	0.0	0.8	0.0
14s		12.3	6.0	0.2	10.2
15p	21.5	2.6	<0.1	0.2	0.0
15s		9.9	5.3	0.9	5.9
21p	24.4	2.6	<0.1	0.1	6.8
21s		16.5	5.1	1.1	0.0
22p	21.9	34.3	1.6	0.3	31.6
22s		26.5	4.9	0.3	5.1

[a] Normal Dutch population; 14 referred for diagnostic workup had normal phenotype
[b] Newborn population of 188 males and 188 females from Albert Einstein Hospital in NY
[c] Randomly selected 7- and 8-year olds from sequential newborn population, consisting of equal numbers of blacks and Caucasians; half with IQ<85; half with IQ≥85
[d] Consecutive newborns, 493 males and 437 females from St. Josephs Hospital, Ontario; parents consisting of 87% Caucasian and 23% black or oriental
[e] Results from 208 normal adults and 141 mentally retarded children of Estonian nationality

Table 3.4 C-band heteromorphism frequencies for chromosomes 1, 9, and 16

Chromosome variant	Holmes [4, 5] Unspec n=20	Müller et al. [8, 9] Mixed newborn n=367	McKenzie and Lubs [10] White n=77	Lubs et al. [3]	
				White n=194	Black N=190
1qh–	10%	0.6%	5%	2%	5%
1gh+	2.50%	8.10%	12%	11%	12%
1qh part inv		1.60%	6%	0.45	0.06
1qh total inv		0	0	0	0
9qh–	7.50%	0.40%	3%	5.00%	3.00%
9qh+	5%	8.00%	10%	7.00%	10.00%
9qh part inv		11.30%	0.45	0.55%	0.45%
9qh total inv	2.50%		1.07	0.13%	1.07%
16qh–		23.60%	39%	35%	39%
16qh+		6.50%	4%	2%	4%
16qh part inv		1.40%	0	0%	0%
16qh total inv	5%	0	0	0%	0%

do not necessarily identify the same variants or reveal any identifiable origin. As already seen, they mainly represent criteria used in different studies to classify variants into groups that seem most similar in size or intensity of staining. Technical variations in staining, in treatment, in degree of chromosome elongation or contraction, and even mutational events can change the appearance of a particular variant within an individual. There are also discrepancies in frequencies because of differences in ethnic origin, age distribution and ascertainment. Nevertheless, the relative frequencies of the most common variants are revealed. Relatively consistent from one study to another is the number of heteromorphisms per individual. By Q and C-banding this number ranges from 4 to 6 with both techniques.

3.1.3 Additional Studies of Racial or Ethnic Differences

Kuleshov and Kulieva [19] studied the overall frequencies of striking variants by G-banding in 6,000 Russian newborns. Variants scored as 1qh+, 9qh+, and 16qh+, were identified as greater than ¼ of the long arm; Gp+ or Dp+, larger than short arm of 18; Ds+ or Gs+, equal to or greater than the thickness of a long arm chromatid; Dss or Gss (double satellites); Yq+ (Y larger than G-group) and Yq– (Y less than the size of G-group); Es+ (satellites on the short arms of chromosomes 17 or 18). The total frequency of these variants combined was 12.8 per 1,000 births. They also determined frequencies in married couples with recurrent abortions and in couples with a history of congenital malformation in offspring (see Sect. 3).

Belloni et al. [20] studied a random sample of newborns in central Italy by C-banding. The frequency of acrocentric variants was higher than reported in other populations. Potluri et al. [21–23] did qualitative analysis of C-band inversion of 1, 9 and 16 studied in 200 infants in New Delhi (100 males and 100 females). Partial (minor) and half inversions of 1 and 9 were observed at modal levels in both sexes. Homozygous size combinations showed higher incidences than heterozygous size combinations for all three chromosomes. However, higher percentages of 1 and 9 inversions were seen in males than females. The percent of size differences between sexes were not significant. Heterozygous inversion combinations of 1 and 9 were more frequent than homozygous combinations in both sexes. No inversion of chromosome 16 was seen. They also observed a significant correlation between C-band size and inversion. The larger the size of the C-band the higher was the incidence of inversion.

3.2 Specialized Banding Studies

Verma, Dosik and Lubs [24] studied variants by QFQ and RFA techniques in 100 white Americans. Six color classes were distinguished by RFA and 5 intensity classes by QFQ. No consistent relationship was seen between color variants by RFA and intensity variants by QFQ. RFA variant frequencies for 13, 14, 15, 21 and 22 were 33.0, 38.0, 28.0, 50 and 24.5%. QFQ variant frequencies for 13, 14, 15, 21 and 22 were 56.5, 10.0, 10.0, and 15.5%. RFA revealed more differences between races than QFQ. In a comparable study of 100 black Americans studied by sequential

QFQ and RFA banding [25], frequencies of QFQ and RFA heteromorphisms were higher in black than in white Americans. No racial difference was noted for chromosome 21 by RFA banding.

References

1. Lubs HA, Ruddle FH (1970) Chromosome abnormalities in the human population: estimation of rates based on New Haven newborn study. Science 169:495–497
2. Lubs HA, Ruddle FH (1971) Chromosome polymorphism in American negro and white populations. Nature 233:134–136
3. Lubs HA, Patil SR, Kimberling WJ, Brown J, Cohen M, Gerald P, Hecht F, Myrianthopoulos N, Summit RL (1977) Q and C-banding polymorphisms in 7 and 8 year old children: racial differences and clinical significance. In: Hook E, Porter I (eds) Population cytogenetic studies in humans. Academic Press, New York, pp 133–159
4. Craig-Holmes AP, Shaw MW (1973) Polymorphism of human C-band heterochromatin. Science 174:702–704
5. Craig-Holms AP, Moore FB, Shaw MW (1973) Polymorphism of human C-band heterochromatin I: frequency of variants. Am J Hum Genet 25:181–192
6. Craig-Holmes AP (1977) C-band polymorphisms in human populations. In: Hook EB, Porter IH (eds) Population cytogenetics. Studies in humans. Academic Press, New York, pp 161–177
7. Geraedts JPM, Pearson PL (1974) Fluorescent chromosome polymorphisms; frequencies and segregation in a Dutch population. Clin Genet 6:247–257
8. Müller HJ, Klinger HP (1975) Chromosome polymorphism in a human newborn population. In: Pearson PL, Lewis KR (eds) Chromosomes today, Vol 5. Wiley, Jerusalem
9. Müller HJ, Klinger HP, Glasser M (1975) Chromosome polymorphism in a human newborn population. II. Potentials of polymorphic chromosome variants for characterizing the ideogram of an individual. Cytogenet Cell Genet 15(4):239–255
10. McKenzie WH, Lubs HA (1975) Human Q and C chromosomal variations: distribution and incidence. Cytogenet Cell Genet 14:97–115
11. Lin CC, Gideon MM, Griffith P, Smink WK, Newton DR, Wilkie L, Sewell LM (1976) Chromosome analysis on 930 consecutive newborn children using quinacrine fluorescent banding technique. Hum Genet 31:315–328
12. Fogle TA, McKenzie WH (1980) Cytogenetic study of a large Black kindred: inversions, heteromorphisms and segregation analysis. Hum Genet 58(3):345–352
13. McKenzie WH, Lubs HA (1975) Human Q and C chromosomal variations: distribution and incidence. Cytogenet Cell Genet 14:97–115
14. Müller HJ, Klinger HP (1975) Chromosome polymorphism in a human newborn population. In: Pearson PL, Lewis KR (eds) Chromosomes today, Vol 5. Wiley, Jerusalem
15. Soudek D, Sroka H (1978) Inversion of "fluorescent" segment in chromosome 3: a polymorphic trait. Hum Genet 44:109–115
16. Mikelsaar A-V, Tuur SJ, Kaosaar ME (1973) Human karyotype polymorphism. I. Routine and fluorescence microscopic investigation of chromosomes in a normal adult population. Humangenetik 20:89–101
17. Mikelsaar AV, Viikmaa MH, Tuur SJ, Koasaar ME (1974) Human chromosome polymorphism. II. The distribution of individuals according to the presence of brilliant bands in chromosomes 3, 4, and 13 in a normal adult population. Humangenetik 23:59–63
18. Mikelsaar AV, Ilus T, Kivi S (1978) Variant chromosome 3 (inv3) in normal newborns and their parents, and in children with mental retardation. Hum Genet 41:109–113

19. Kuleshov NP, Kulieva LM (1979) Frequency of chromosome variants in human populations Chastota khromosmnykh variantov v populiatsiiakh cheloveka. Genetika 15:745–751
20. Belloni G, Benincasa A, Bosi A, de Capoa A, Di Castro M, Ferraro M, Lombardi D, Mostacci C, Pelliccia F, Prantera G et al (1983) Screening for cytogenetic polymorphisms in a random sample of live born infants from Italian population. Acta Anthropogenet 7:205–217
21. Potluri VR, Singh IP, Bhasin MK (1985) Chromosomal heteromorphisms in Delhi infants. III. Qualitative analysis of C-band inversion heteromorphisms of chromosomes 1, 9 and 16. J Heredity 76:55–58
22. Potluri VR, Singh IP, Bhasin MK (1985) Human chromosomal heteromorphisms in Delhi newborns. II. Analysis of C-band size heteromorphisms in chromosomes 1, 9 and 16. Hum Heredity 35:333–338
23. Potluri VR, Singh IP, Bhasin MK (1985) Human chromosomal heteromorphisms in Delhi newborns. VI. Inter-relationship between C-band size and inversion heteromorphisms. Cytobios 44:149–152
24. Verma RS, Dosik H, Lubs HA (1977) Frequency of RFA colour polymorphisms of human acrocentric chromosomes in Caucasians: inter-relationship with QFQ polymorphisms. Ann Hum Genet 41:257–267
25. Verma RS, Dosik H (1981) Human chromosomal heteromorphisms in American Blacks. III. Evidence for racial differences in RFA color and QFQ intensity heteromorphisms. Hum Genet 56:329–337

Chapter 4
Clinical Populations

Persistent notions that striking chromosomal heteromorphisms are directly associated with clinical anomalies or have some indirect effect on the frequencies of major chromosome abnormalities or spontaneous miscarriages, have been the topics of numerous studies. Early studies [1–4] suggested roles of striking variants in mental retardation, autism, behavior disorders and congenital anomalies. In general, such studies were small and most were not substantiated. Soudek et al. [5] gave four attributes of heteromorphisms: (1) they contained repetitive satellite DNA; (2) they were inherited; (3) there were no syndromes associated; and, (4) if phenotypic effects were associated, they were most likely due to an indirect selective effect. Barlow [6] suggested that extra heterochromatin might affect birth weight, body weight, immunoglobulin levels and cell growth. He also pointed out the difficulties technical variables make in comparing different studies. Similar points were made by Maes et al. [7]. Generally, there was no definitive evidence for a direct role of variations in size of the common heteromorphisms on phenotype or in mental retardation.

4.1 Spontaneous Abortions and Reproductive Failure

Numerous studies attempted to show increased rates of spontaneous abortion or reproductive failure associated with striking variants such as Yqh+, 9qh+, 9qh− and inv(9qh), or with the amount of heterochromatin overall [7–19]. Again, such studies were generally inconclusive and conflicting. In studies of heterochromatic variants, Tsvetkova and Iankova [13] found no differences in frequencies of routine variants in couples with reproductive failure and normal couples. Maes et al. [7] also found no effects of heterochromatin length in Caucasian couples with recurrent early abortion. Kuleshov and Kulieva [14] studied frequencies of extreme variants in over 400 married couples with recurrent abortions and couples with a history of congenital malformations in offspring. Excluding major chromosome abnormalities, 14.6% of couples in the first group had extreme variants vs. 13.3% of couples in the second group. Neither frequency was impressively different from an overall

H. E. Wyandt, V. S. Tonk, *Human Chromosome Variation: Heteromorphism and Polymorphism,* DOI 10.1007/978-94-007-0896-9_4,

frequency of 12.6% in 6,000 newborns. Interestingly, only 7.2% of patients with Down syndrome had extreme variants.

In a critical review of the literature on association of heterochromatic variants with reproductive failure manifesting as infertility or recurrent spontaneous abortions, Bobrow [15] found that differences in methods made data difficult to interpret. However, the majority of evidence weighed against any significant effect of autosomal variants. Although reports for the Y were conflicting, he concluded that the effects of Yq− in infertility and Yq+ in recurrent abortion were mainly positive. Rodriguez-Gomez et al. [16] subsequently examined heterochromatin length of 1, 9, 16 and Y on reproductive wastage in couples with recurrent miscarriages versus normal couples and concluded that neither Y nor autosomal C-band size was directly related. Others [17–19] also found no effects of the amount of autosomal heterochromatin or the length of Y heterochromatin on repeat abortions. Buretic-Tomljanovich et al. [18] did find an increase in heterochromatin of chromosome 16 in couples with a stillborn or stillborn malformed child.

4.2 Non-disjunction

A persistent idea was that of the influence of heteromorphic regions themselves on non-disjunction. Jackson-Cook et al. [20] studied "double NORs" (dNOR) in 50 families having a child with trisomy 21. Of 41 informative sets of parents, 13 that were shown to contribute the extra chromosome had a dNOR. Only one dNOR was found in the 41 normal spouses and none were found in 50 additional normal control subjects, strongly suggesting a role of dNORs on non-disjunction in Down syndrome. Green et al. [21] subsequently studied NOR size and frequency in 43 parents of Down syndrome children and 39 controls. The risk for having a child with trisomy 21 correlated with a higher frequency of associations and with number of NORs per cell, but specific types of NOR variants such as large NORs or dNORs appeared not to be correlated. Studies by other groups [22–24] have also not been able to confirm a role of dNORs in Down syndrome.

Several problems in determining whether NORs play a role in non-disjunction may be technical. Although variation in size or number of NORs is often regarded to be heritable and fixed [25, 26], NOR staining was early suggested to reflect NOR activity [27–29], and the size and number of NORs was shown to correlate with the frequency of acrocentric associations [30–32]. Balicek and colleagues [33, 34] studied the intercalary fine structure of multiple satellites on acrocentrics. They showed the light staining regions were heat resistant and stained heavily with the RGH (R-banding by heat, Giemsa) technique, indicating these regions were GC-rich. Such regions, which become extremely under-condensed when 5-azacytidine is added to cultures, showed variably positive silver-staining anywhere along their length. Regions that were negatively silver stained but RGH-positive were interpreted to be "inactive NOR sites", whereas silver positive regions were felt to be "active NOR sites". Perez-Castillo et al. [35] reported a patient with a very long homogeneous short arm on a chromosome 15 showing up to four secondary constrictions in other sibs. The region stained similarly by Q- and G-bands in the four carriers, but stained less intensely in the proband by C-banding. DAPI staining showed only a proximal intense band. Silver staining revealed from 0 to 4 silver-positive regions. If only a single NOR were present, it was usually the distal one. The distribution and number of NOR regions on different acrocentric chromosomes varied in each carrier. However, the mean number for all chromosomes was similar for the different carriers, suggesting an optimal threshold level of NOR activity. The exception was the index case that showed a mean value significantly higher than the other four.

A variety of mechanisms have been proposed for the cause of nondisjunction leading to trisomy 21 in Down's syndrome. Aside from a parental age effect and determination of parental and meiotic origin, mechanisms causing non-disjunction have remained elusive. Familial or genetic risks for increased rates of trisomy have been proposed [36], while others [37, 38] have suggested different categories of risk based on DNA polymorphism studies. More recent evidence implicates position and number of meiotic cross-over events in non-disjunction [39, 40].

4.3 Satellite Association

Closely related to the issue of NORs in non-disjunction is that of satellite association. Satellite association has been speculated to increase the risk for non-disjunction and present a mechanism for generating Robertsonian translocations. Several studies have suggested that unusually high rates of satellite association in patients with reproductive difficulties while others do not [41–43]. Taken together, the current literature does not establish that satellite association has any effect upon reproductive failure.

4.4 Cancer

A significant number of studies have suggested an association of striking heteromorphisms with a surprisingly wide range of cancers and leukemia (Table 4.1) [44]. Atkin and Baker [45] first suggested susceptibility to malignancy associated with heteromorphism in chromosome 1. Subsequent observations were reported for chromosomes 1, 9 and 16 and the Y and include observations of increased or decreased length, striking size difference between homologs (asymmetry), and pericentric inversions in heterochromatic regions. A few studies showed no differences between cancer patients and controls [46–48] or among tissues from the same individual, indicating that the observed heteromorphic variants were constitutional in nature [49–51]. An apparently acquired common pericentric inversion of chromosome 9, i.e. inv(9)(p11q13), was initially reported in bone marrow, but not in PHA-stimulated blood from a patient with essential thrombocytopenia [52]. Betz

Table 4.1 Heteromorphisms associated with various forms of cancer and leukemia. (Adapted from Wyandt and Patil [44])

Disease	Chromosome and type
A. Solid tumors	
Bladder carcinoma	1qh+; inv(1qh); structural abnormalities of 1; 21nor+; C-band size and location
Breast	1qh variants; 1qh+, 9qh+, 16qh+, inv(1), inv(9qh)
Ovarian	Structural abnormalities of 1, inv(1qh), 1qh asymmetry
Cervical cancer and dysplasia	1qh variants, partial inversions
Cervical and uterine	Partial inversions
Child malignancy	9qh+; 1,9 and 16qh asymmetry
Colon and rectal	Inv(9qh), inv(1qh)
Endometrial cancer	Increased heterochromatin length
Malignancy (non-spec.)	1qh and 9qh variants; breakage; inv(1qh); 1, 9 and 16qh asymmetry
Pituitary adenoma	Inv(9qh)
Prostate cancer, <age 70	1qh+ and 16qh+
Squamous cell carcinoma	Structural abnormalities of 9qh
Testicular	9qh++, 16qh++ (borderline signif.)
B. Hematological	
NHL	Inv(1qh), 1qh−, 9qh−
CML, PV and MDS	9qh−(+other); 9qh+(−other); 1qh+; 1qh−; 1, 9 and 16qh, asymmetry; inv(9qh)
Pre-leukemia	1qh+, 1qh−
Leukemia	Yqh+
AML	Excessive sym 9qh; 1qh, 1qh−
ALL and childhood ALL	1, 9 and 16qh−

et al. [53] reported two additional cases of acquired inv(9)(p11q13), in a patient with acute myeloid leukemia and in a patient with severe anemia, respectively. The significance of the inversions in the pathogenesis of these cases is speculative.

Despite the volume of reports, the role, if any, of constitutive heterochromatin variation in malignancy remains unclear. Association of various cancers and leukemias with higher heteromorphism frequencies, especially asymmetry (striking differences between homologs) or excessive symmetry (suggesting identical homologs) is intriguing. Cohorts of cancer patients have been small and few studies have rigorously matched patients and controls for age, ethnicity, race, etc. As aptly pointed out by Erdtmann [54], technical factors such as quality of preparations, criteria for scoring variants and variations in methodology do not allow direct comparisons of results from different studies. Also, few studies have tried to differentiate variations that are *acquired* from those that are *constitutional* in nature, a distinction that may be important in determining whether an observed difference is causal or a result of the neoplastic process. Finally, few studies have looked for *mosaicism* in cancer patients vs. normal controls. In a study of three cancer families, Doneda et al. [55] found C-band length mosaicism for chromosome 1 in lymphocytes in six cancer patients and one normal individual.

Qu et al. [56] reported frequent hypomethylation of pericentromeric DNA in chromosomes 1 and 16 in Wilms' tumors. They found a significant relationship between the loss of 16q and hypomethylation of 16qh satellite 2 DNA, suggesting that hypomethylation is causally involved. However, an argument against involvement of hypomethylation of heterochromatin in cancer is the notable example of ICF syndrome, a rare autosomal recessive genetic condition that has the characteristic features of immunodeficiency, centromeric heterochromatin instability, and facial anomalies, without associated cancer. Frequent rosette formation and multiple copies of chromosome arms joined at their h regions [57] may indicate defective DNA methylation. Classical satellite 2 DNA in the pericentromeric regions of chromosomes 1 and 16 in leukocytes from ICF patients is hypomethylated compared to leukocytes from normal individuals [58, 59].

Finally, not all apparent variants involve only heterochromatin. Rearrangements in the pericentromeric region of chromosome 1 or 16, for example, frequent in several types of cancers, are known to involve particular oncogenes that are close to the pericentromeric regions [60, 61]. Inversions or insertions of these genes into heterochromatin regions could conceivably play a role in turning such genes on or off by position effect.

References

1. Lubs HA, Patil SR, Kimberling WJ, Brown J, Cohen M, Gerald P, Hecht F, Myrianthopoulos N, Summit RL (1977) Q and C-banding polymorphisms in 7 and 8 year old children: racial differences and clinical significance. In: Hook E, Porter I (eds) Population cytogenetic studies in humans. Academic Press, New York, pp 133–159
2. Tharapel AT, Summitt RL (1978) Minor chromosome variations and selected heteromorphisms in 200 unclassifiable mentally retarded patients and 200 normal controls. Hum Genet 41:121–130
3. Funderburk SJ, Guthrie D, Lind RC, Müller HM, Sparkes RS, Westlake JR (1978) Minor chromosome variants in child psychiatric patients. Am J Med Genet 1:301–308
4. Matsuura JS, Mayer M, Jacobs PA (1979) A cytogenetic survey of an institution for the mentally retarded. III. Q-band chromosome heteromorphisms. Hum Genet 523:203–210
5. Soudek D, Sroka H (1979) Chromosomal variants in mentally retarded and normal men. Clin Genet 16:109–116
6. Barlow P (1973) The influence of inactive chromosomes on human development. Hum Genet 17:105–136
7. Maes A, Stassen C, Hens L, Vamos E, Kirsch-Volders M, Lauwers MC, Defrise-Gussenhoven E, Susanne C (1983) C heterochromatin variation in couples with recurrent early abortions. J Med Genet 20:350–356
8. Patil SR, Lubs HA (1977) A possible association of long Y chromosomes and fetal loss. Hum Genet 35:233–235
9. Genest P (1979) Chromosome variants and abnormalities in 51 married couples with repeated spontaneous abortions. Clin Genet 16:387–389
10. Ford JH, Lester P (1978) Chromosomal variants and nondisjunction. Cytogenet Cell Genet 21:300–303
11. Eiben B, Leipoldt M, Rammelsberg O, Krause W, Engel W (1987) High incidence of minor chromosomal variants in teratozoospermic males. Andrologia 19:684–687

12. Del Porto G, D'Alessandro E, Grammatico P, Coghi IM, DeSanctis S, Giambenedetti M, Vaccarella C, Fabi R, Marciano MF, Nicotra M (1993) Chromosome heteromorphisms and early recurrent abortions. Hum Reprod 8:755–758
13. Tsvetkova TG, Iankova MF (1979) Human chromosome polymorphism and disordered reproductive function. I. Routine chromosome variants [in Russian]. Khromosomnyi polimorfism i narushenie reproduktivnoi funktsii u cheloveka. Soobshchenue I. Rutinnye varianty khromosom. Genetika 15:1858–1869
14. Kuleshov NP, Kulieva LM (1979) Frequency of chromosome variants in human populations [in Russian]. Chastota khromosmnykh variantov v populiatsiiakh cheloveka. Genetika 15:745–751
15. Bobrow M (1985) Heterochromatic chromosome variation and reproductive failure. Exp Clin Immunogenet 2:97–105 (Review)
16. Rodriguez-Gomez MT, Martin-Sempere MJ, Abrisqueta JA (1987) C-band length variability and reproductive wastage. Hum Genet 75:56–61
17. Kruminia AR, Kroshkina VG, Voskoboinik NI, Reshetnikov AN (1987) Quantitative analysis of C segments of chromosomes 1,9,16 and Y in couples with reproductive disorders [Russian]. Genetika 23:540–543.
18. Buretic-Tomljanovic A, Rodojcic Badovinac A, Vlastelic I, Randic LJ (1997) Quantiative analysis of constitutive heterochromatin in couples with fetal wastage. Am J Reprod Immun 38:201–204
19. Podugol'nikova OA, Solonichenko VG (1994) The C heterochromatin of chromosomes 1, 9, 16 and Y in patients with Noonan's syndrome [in Russian]. Tsitol Genet 28:85–88
20. Jackson-Cook CK, Flannery DB, Corey LA, Nance WE, Brown JA (1985) Nucleolar organizer region variants as a risk factor for Down syndrome. Am J Hum Genet 37:1049–1061
21. Green JE, Rosenbaum KN, Rapoport SI, Schapiro MB, White BJ (1989) Variant nucleolar organizing regions and the risk of Down syndrome. Clin Genet 35:243–250
22. Serra A, Bova R (1990) Acrocentric chromosome double NOR is not a risk factor for Down syndrome. Am J Med Genet Suppl 7:169–174
23. Sheng WW, Perng CK, Wuu KD (1992) Double NOR is not a good indicator of risk for Down syndrome. Jpn J Hum Genet 37:151–155
24. Hassold T, Jacobs PA, Pettay D (1987) Analysis of nucleolar organizing regions in parents of trisomic spontaneous abortions. Hum Genet 76(4):381–384
25. Mikelsaar AV, Schwarzacher HG, Schnedl W, Wagenbichler P (1977) Inheritance of Ag-stainability of nucleolus organizer regions. Investigations in 7 families with trisomy 21. Hum Genet 38:183–188
26. Markovic VD, Worton RG, Berg JM (1978) Evidence for the inheritance of silver-stained nucleolus organizer regions. Hum Genet 41(2):181–187
27. Dev VG, Byrne J, Bunch G (1979) Partial translocation of NOR and its activity in a balanced carrier and in her cri-du-chat fetus. Hum Genet 51:277–280
28. Miller DA, Dev VG, Tantravahi R, Croce CM, Miller OJ (1978) Human tumor and rodent-human hybrid cells with an increased number of active human NORs. Cytogenet Cell Genet 2:33–41
29. Lau Y-F, Wertelecki W, Pfeiffer RA, Arrighi FE (1979) Cytological analyses of a 14p+ variant by means of N-banding and combinations of silver staining and chromosome bandings. Hum Genet 46:75–82
30. Bernstein R, Dawson B, Griffiths J (1981) Human inherited marker chromosome 22 short-arm enlargement: investigation of rDNA gene multiplicity, Ag-band size, and acrocentric association. Hum Genet 58:135–139
31. Warburton D, Atwood KC, Henderson AS (1976) Variation in the number of genes for rRNA among human acrocentric chromosomes: correlation with frequency of satellite association. Cytogenet Cell Genet 17:221–230
32. Miller DA, Tantravahi R, Dev VG, Miller OJ (1977) Frequency of satellite association of human chromosomes is correlated with amount of Ag-staining of the nucleolus organizer region. Am J Hum Genet 29:490–502

33. Balicek P, Zizka J (1980) Intercalar satellites of human acrocentric chromosomes as a cytological manifestation of polymorphisms in GC-rich material. Hum Genet 54:343–347
34. Balicek P, Zizka J, Skalska H (1982) RGH-band polymorphism of the short arms of human acrocentric chromosomes and relationship of variants to satellite association. Hum Genet 62:237–239
35. Perez-Castillo A, Martin-Lucas MA, Abrisqueta JA (1986) New insights into the effects of extra nucleolus organizer regions. Hum Genet 72:80–82
36. Hobbs CA, Sherman SL, Yi P, Hopkins SE, Torks CP, Hine RJ, Pogribna M, Rozen R, James SJ (2000) Polymorphisms in genes involved in folate metabolism as maternal risk factors for Down syndrome. Am J Hum Genet 67:623–630
37. Pangalos CG, Talbot CC Jr, Lewis JG, Adelsberger PA, Petersen MB, Serre JL, Rethore MO, de Blois MC, Parent P, Schinzel AA et al (1992) DNA polymorphism analysis in families with recurrence of free trisomy 21. Am J Hum Genet 51:1015–1027
38. Pangalos C, Avramopoulos D, Blouin JL, Raoul O, deBlois MC, Prieur M, Schinzel AA, Gika M, Abazis D, Antonarakis SE (1994) Understanding the mechanism(s) of mosaic trisomy 21 by using DNA polymorphism analysis. Am J Hum Genet 54:473–481
39. Warren AC, Chakravarti A, Wong C, Slaugenhaupt SA, Halloran SL, Watkins PC, Metaxotou C, Antonarakis SE (1987) Evidence for reduced recombination on the nondisjoined chromosomes 21 in Down syndrome. Science 237:652–654
40. Sherman SL, Takaesu N, Freeman SB, Grantham M, Phillips C, Blackston RD, Jacobs PA, Cockwell AE, Freeman V, Uchida I et al (1991) Trisomy 21: association between reduced recombination and nondisjunction. Am J Hum Genet 49:608–620
41. Yasseen AA, Aunuiz AF (2002) High frequency of satellite association in metaphases of infertile male patients. Saudi Med J 23:427–431
42. Jacobs PA, Mayer M (1981) The origin of human trisomy: a study of heteromorphisms and satellite associations. Ann Hum Genet 45:357–365
43. Kovaleva NV, Butomo IV, Novikova I (1993) Acrocentric chromosomal associations in the families of children with Down's disease. Tsitologia 35:33–43
44. Wyandt HE, Patil RS (2004) Heteromorphisms in clinical populations. In: Wyandt HE, Tonk VS (eds) Atlas of human chromosome heteromorphisms. Kluwer, Dordrecht, pp 47–62
45. Atkin NB (1977) Chromosome 1 heteromorphism in patients with malignant disease: a constitutional marker for a high risk group? Br Med J 1:358
46. Lundgren R, Berger R, Kistoffersson U (1991) Constitutive heterochromatin C-band polymorphism in prostatic cancer. Cancer Genet Cytogenet 51:57–62
47. Kivi S, Mikelsaar AV (1980) Q- and C-band polymorphisms in patients with ovarian or breast carcinoma. Hum Genet 56:111–114
48. Kivi S, Mikelsaar AV (1987) C-band polymorphisms in lymphocytes of patients with ovarian or breast cancer. Cancer Genet Cytogenet 28:77–85
49. Heneen WK, Habib ZA, Rohme D (1980) Heteromorphism of constitutive heterochromatin in carcinoma and dysplasia of the uterine cervix. Eur J Obstet Gynecol Reprod Biol 10:173–182
50. Labal de Vinuesa M, Larripa I, Mudry de Pargament M, Brieux de Salum S (1984) Heterochromatic variants and their association with neoplasias. I. Chronic and acute leukemia. Cancer Genet Cytogenet 13:297–302
51. Sadamori N, Sandberg AA (1983) The clinical and cytogenetic significance of C-banding on chromosome #9 in patients with Ph1-positive chronic myeloid leukemia. Cancer Genet Cytogenet 8:235–241
52. Wan TS, Ma SK, Chan LC (2000) Acquired pericentric inversion of chromosome 9 in essential thrombocythemia. Hum Genet 106:669–670
53. Betz JL, Behairy AS, Rabionet P, Tirtorahardjo B, Moore MW, Cotter PD (2005) Acquired inv(9): what is its significance? Cancer Genet Cytogenet 160:76–78
54. Erdtmann B (1982) Aspects of evaluation, significance and evolution of human C-band heteromorphism. Hum Genet 61:281–294
55. Doneda L, Conti AF, Gualandri V, Larizza L (1987) Mosaicism in the C-banded region of chromosome 1 in cancer families. Cancer Genet Cytogenet 27:261–268

56. Qu GZ, Grundy PE, Narayan A, Ehrlich M (1999) Frequent hypomethylation in Wilms tumors of pericentromeric DNA in chromosomes 1 and 16. Cancer Genet Cytogenet 109:34–39
57. Sawyer JR, Swanson CM, Wheeler G, Cunniff C (1995) Chromosome instability in ICF syndrome: formation of micronuclei from multibranched chromomosomes 1 demonstrated by fluorescence *in situ* hybridization. Am J Med Genet 56:203–209
58. Jeanpierre M, Turleau C, Aurias A, Prieur M, Ledeist F, Fischer A, Viegas-Pequignot E (1993) An embryonic-like methylation pattern of classical satellite DNA is observed in ICF syndrome. Hum Mol Genet 2:731–735
59. Ji W, Hernandez R, Zhang XY, Qu GZ, Frady A, Varela M, Ehrlich M (1997) DNA demethylation and pericentromeric rearrangements of chromosome 1. Mut Res 379:33–41
60. Tse W, Zhu W, Chen HS, Cohen A (1995) A novel gene, AF1q, fused to MLL in t(1;11)(q21;q23), is specifically expressed in leukemic and immature hematopoietic cells. Blood 85:650–656
61. Mugneret F, Dastugue N, Favre B, Sidaner I, Salles B, Huguet-Rigal F, Solary E (1995) Der(16)t(1;16)(q11;q11) in myelodysplastic syndromes: a new non-random abnormality characterized by cytogenetic and fluorescence *in situ* hybridization studies. Brit J Haematol 90:119–124

Chapter 5
Euchromatic Variants

A new class of variants that does not fit in with the usual perception of heteromorphism involve so-called "euchromatic variants", regions that are C-band negative and not generally anticipated to be variable in size or staining because they presumably contain genetic material. Euchromatic variants have become increasingly important at the molecular level where deletions and duplications occur as normal variants throughout the genome (see Copy Number Variants, Part IV, Chapter 32). However, examples of such variants detectable at the microscope level are rare and are certainly appropriate subjects for critical scrutiny when they are observed. Euchromatic duplications and deletions (Table 5.1) originally compiled by Jalal and Ketterling [1] included both light (G-negative) and dark (G-positive) bands identified by G-banding that seemed to be phenotypically neutral. They emphasized that all euchromatic duplications and deletions should be interpreted with caution. In order for a euchromatic deletion or duplication to be regarded as a normal variant, they indicated the following criteria should be met. "It should have been: (1) reported in a relatively large number of individuals; (2) been passed on from parents to children; (3) associated with a normal phenotype; (4) the identity of the extra or missing chromatin confirmed by chromosome banding and/or molecular/molecular cytogenetic procedures." Also, emphasized for euchromatic duplications, was the necessity to rule out moderately or highly repetitive DNA sequences that are C-band negative, such as alpha satellite, satellite III and ribosomal DNA. Further cited, was evidence of the presence and amplification of pseudogene cassettes in 16p11.2 that include pseudogenes for immunoglobulin heavy chain from 14q32, myosin heavy chain from Xq28, minisatellite sequences from the telomeric region of 6p [2], and reports of paralogous segments of Xq28 found near the centromeres of 2p, 10p, 16p and 22q [3, 4]. It has since has become increasing evident that microduplications, deletions and amplifications involving microsatellites and pseudogenes occur throughout the genome and that such sequences at telomeric, centromeric, neocentromeric, and interstitial fragile sites have significant roles in chromosome rearrangement, in gene expression and in chromosome evolution. Therefore, a goal to characterize unusual variants with greater precision at the molecular level is certain to be a worthwhile and appropriate mission of the newer array technologies discussed in Part IV, Chapter 32.

H. E. Wyandt, V. S. Tonk, *Human Chromosome Variation: Heteromorphism and Polymorphism,* DOI 10.1007/978-94-007-0896-9_5,

Table 5.1 Euchromatic variants

Chr	Duplications	Deletions	Reference(s)
1	1p21-p31		[5]
	1q42.11-q42.12		[6]
2		2q14.1	[7]
3		3p25.3	[8]
4	4q35		Fig. 9.1d
5		5p14	[9–12]
	5q15-q21		[13]
6		6q23.1-q24.2	[14–16]
8	8p23.1		[17–19]
		8q24.13-q24.22	[20–22]
9	9p12?		[23]
	9q13-q21		[24–26]
11		11p12	[27]
13		13q21	[28]
15	15q12.2-q13.1		[29–37]
16	16p11.1-p12		[38–42]
		16q21	[9, 43–46]
18	18p+		[47–52]
X		Xq26	[53]

References

1. Jalal SM (2004) Ketterling. Euchromatic Variants. In: Wyandt HE, Tonk VS (eds) Atlas of human chromosome heteromorphisms. Kluwer, Dordrecht, pp 75–95
2. Barber JCK, Reed CJ, Dahoun SP, Joyce CA (1999) Applification of a pseudogene cassette underlies euchromatic variation of 16p at the cytogenetic level. Hum Genet 104:211–218
3. Eichler EE, Lu F, Shen Y, Antonacci R, Jurecic V, Doggett NA, Mayzis RK, Baldini A, Gibbs RA, Nelson DL (1996) Duplication of a gene-rich cluster between 16p11.1 and Xq28: a novel pericentromeric-directed mechanism for paralogous genome evaluation. Hum Mod Genet 5:899–913
4. Eichler EE, Budarf ML, Rocchi M, Deaven LL, Doggett NA, Baldini A, Nelson DL, Mohrenweiser HW (1997) Interchromosomal duplications of the adrenoleukodystrophy locus: a phenomenon of pericentromeric plasticity. Hum Mol Genet 6:991–1002
5. Zazlov AL, Blumenthal D, Fox JE, Thomson KA, Segraves R, Weinstein ME (1993) A rare inherited euchromatic heteromorphism on chromosome 1. Prenat Diagn 13:569–573
6. Bortotto L, Piovan E, Furlan R, Rivera H, Zuffardi O (1990) Chromosome imbalance, normal phenotype, and imprinting. J Med Genet 27:582–587
7. Sumption ND, Barber JCK (2001) A transmitted deletion of 2q13 to 2q14.1 causes no phenotypic abnormalities. J Med Genet 38:125–127
8. Knight LA, Yong MH, Tan M, Ng ISL (1995) Del(3)(p25.3) without phenotypic effect. J Med Genet 32:994–995
9. Hand JL, Michels VV, Marinello MJ, Ketterling RP, Jalal SM (2000) Inherited interstitial deletion of chromosomes 5p and 16q without apparent phenotypic effect: further confirmation. Prenat Diagn 20:144–148

10. Walker JL, Blank CE, Smith BAM (1984) Interstitial deletion of the short arm of chromosome 5 in a mother and three children. J Med Gen 21(6):465–467
11. Overhauser J, Golbus MS, Schonberg SA, Wasmuth JJ (1986) Molecular analysis of an unbalanced deletion of the short arm of chromosome 5 that produces no phenotype. Am J Hum Genet 39:1–10
12. Keppen LD, Gollin SM, Edwards D, Sawyer J, Wilson W, Overhauser J (1992) Clinical phenotype and molecular analysis of a three-generation family with an interstitial deletion of the short arm of chromosome 5. Am J Med Genet 44:356–360
13. Shuan-Yow L, Gibson LH, Gomez K, Pober BR, Yang-Feng TL (1998) Familial dup(5)(q15q21) associated with normal and abnormal phenotypes. Am J Med Genet 75:75–7
14. Kumar A, Cassidy SB, Romero L, Schwartz S (1999) Molecular cytogenetics of a de novo interstitial deletion of chromosome arm 6q in a developmentally normal girl. Am J Med Genet 86:227–231
15. Matkins SV, Meyer JE, Berry AC (1987) A child with partial monosomy 6q secondary to a maternal direct insertional event. J Med Genet 24:227–229
16. Christian SL, Rich BH, Loebl C, Israel J, Vasa R, Kittikamron K, Spiro R, Rosenfield R, Ledbetter DH (1999) Significance of genetic testing for paternal uniparental disomy of chromosome 6 in neonatal diabetes mellitus. J Pediatr 134:42–46
17. Barber JCK, Joyce CA, Collinson MN, Nicholson JC, Wilatt LR, Dyson HM, Bateman MS, Green AJ, Yates JRW, Dennis RD (1998) Duplication of 8p32.1: a cytogenetic anomaly with no established clinical significance. J Med Genet 35:491–496
18. Engelen JJM, Moog U, Evers JLH, Dassen H, Albrechts JCM, Hamers AJH (2000) Duplication of chromosome region 8p23.1→p23.3. Am J Med Genet 91:18–21
19. Williams L, Larkin S, Roberts E, Davisson EV (1996) Two further cases of variations in banand 8p23.1. Not always a benign variant. J Med Genet 33:522
20. Ludecke H, Johnson C, Wagner M et al (1991) Molecular definition of the shortest region of deletion overlap in the Langer-Giedion syndrome. Am J Hum Genet 49:1197–1206
21. Batanian JR, Morris K, Ma E, Huang Y, McComb J (2001) Familial deletion of (8)(q24.1q24.22) associated with a normal phenotype. Clin Genet 60:371–373
22. Fennel SJ, Benson JW, Kindley AD, Schwarz MJ, Czepulkowski B (1989) Partial deletion 8q without Langer-Giedion syndrome: a recognizable syndrome. J Med Genet 26(3):167–171
23. Sutherland GR, Eyre H (1981) Two unusual G-band variants of the short arm of chromosome 9. Clin Genet 19:331–334
24. Wang JC, Miller WA (1994) Molecular cytogenetic characterization of two types of chromosome 9 variants. Cytogenet Cell Genet 67:190–192
25. Jalal SM, Kukolich MK, Garcia M, Day DW (1990) Euchromatic 9q+ heteromorphism in a family. Am J Hum Genet 37:155–156
26. Knight LA, Soon GM, Tan M (1993) Extra G positive band on the long arm of chromosome 9. J Med Genet 30:613
27. Barber JCK, Mahl H, Portch J, Crawfurd MD (1991) Interstitial deletions without phenotypic effect: prenatal diagnosis of a new family and brief review. Prenat Diagn 11:411–416
28. Coutuier J, Morichon-Delvallez N, Dutrillaux B (1986) Deletion of band 13q21 is compatible with normal phenotype. Hum Genet 70:87–91
29. Brookwell R, Veleba A (1994) Proximal 15q variant with normal phenotype in three unrelated individuals. Clin Genet 31:311–314
30. Ludowese CJ, Thompson KJ, Sekon GS, Pauli RM (1991) Absence of predictable phenotypic expression in proximal 15q duplications. Clin Genet 40:194–201
31. Jalal SM, Persons OL, Dewald GW (1994) Form of 15q proximal duplication appears to be a normal euchromatic variant. Am J Med Genet 52:495–497
32. Mao R, Jalal SM, Snow K, Michels VV, Szabo SM, Babovic-Vuksanovic D (2000) Characteristics of two cases with dup(15)(q11.2–12): one of maternal and one of paternal origin. Genet In Med 2:131–135
33. Riordan D, Dawson AJ (1998) The evaluation of proximal 15q duplications by FISH. Clin Genet 54:517–521

34. Browne CE, Dennis NR, Maher E et al (1997) Inherited interstitial duplications of proximal 15q: genotype-phenotype correlations. Am J Med Genet 61:1342–52
35. Bolton PF, Dennis NR, Browne CE, Thomas NS, Veltman MWM, Thompson RJ, Jacobs P (2001) The phenotypic manifestations of interstitial duplications of proximal 15q with special reference to the autistic spectrum disorders. Am J Med Genet (Neuro Genet) 105:675–85
36. Yardin C, Esclaire F, Laroche C, Terro F, Barthe D, Bonnefont J-P, Gilbert B (2002) Should the chromosome region 15q11q13 be systematically tested by FISH in the case of an autistic-like syndrome? Clin Genet 61:310–13
37. Moeschler JB, Mohandas TK, Hawk AB, Knoll W (2002). Research Letter. Estimate of the prevalence of proximal 15q duplication syndrome. Am J Med Genet 111:440–42
38. Thompson PW, Roberts SH (1987) A new variant of chromosome 16. Hum Genet 76:100–101
39. Thompson PW, Roberts SH, Rees SM (1990) Replication studies in the 16p+ variant. Hum Genet 84:371–372
40. Pinel I, de Bustamante AD, Urioste M, Felix V, Ureta A, Martinez-Frias ML (1988) An unusual variant of chromosome 16. Two new cases. Hum Genet 80:194
41. Jalal SM, Schneider NR, Kukolich MK, Wilson GN (1990) Euchromatic 16p+ heteromorphism: first report in North America. Am J Med Genet 37:548–550
42. Bryke CR, Breg WR, Potluri VR, Yang-Feng TL (1990) Duplication of euchromatin without phenotypic effects: a variant of chromosome 16. Am J Med Genet 36:43–44
43. Verma RS, Kleyman SM, Conte RA (1997) Variant euchromatic band within 16q12.1. Clin Genet 52:446–447
44. Naritomi K, Shiroma N, Izumikawa Y, Sameshima K, Ohdo S, Hirayama K (1988) 16q21 is critical for 16q deletion syndrome. Clin Genet 33:372–375
45. Witt DR, Lew SP, Mann J (1988) Heritable deletion of band 16q21 with normal phenotype: relationship to late replicating DNA. Am J Hum Genet 43:A127
46. Callen DF, Eyre H, Lane S, Shen Y, Hansmann I, Spinner N, Zackai E, McDonald-McGinn D, Schuffenhauer S, Wauters J, Van Thienen M-N, Van Roy B, Sutherland GR, Haan EA (1993) High resolution mapping of interstitial long arm deletions of chromosome 16: relationship to phenotype. J Med Genet 30:828–832
47. Wolff DJ, Raffel LJ, Ferre MM, Schwartz S (1991) Prenatal ascertainment of an inherited dup(18p) associated with an apparently normal phenotype. Am J Med Genet 41:319–321
48. Moog U, Engelen JJ, de Die-Smulders CE, Albrechts JC, Loneus WH, Haagen AA, Raven EJ, Hamers AJ (1994) Partial trisomy of the short arm of chromosome 18 due to inversion duplication and direct duplication. Clin Genet 46:423–429
49. Abeliovich D, Dagan J, Levy A, Steinberg A, Zlotogora J (1993) Isochromo-some 18p in a mother and her child. Am J Med Genet 46:392–393
50. Johnasson B, Mertens F, Palm L, Englesson I, Kristofferson U (1988) Duplication 18p with mild influence on phenotype. Am J Med Genet 29:871–874
51. Singer TS, Kohn G, Yatziv S (1990) Tetrasomy 18p in a child with trisomy 18 phenotype. Am J Med Genet 36:144–147
52. Pinto MR, Silva ML, Ribeiro MC, Pina R (1998) Prenatal diagnosis of mosaicism for tetrasomy 18p: cytogenetic, FISH and morphological findings. Prenat Diagn 18:1095–1097
53. Taysi K (1983) Del(X) (q26) in a phenotypically normal woman and her daughter who also has trisomy 21. Am J Med Genet 14:367–372

Part II
Chromosome Heteromorphism (Summaries)

The following summaries include examples of human chromosome heteromorphisms that have been compiled by us, contributed by individual investigators specifically for this volume, or have been reprinted with permission from various published sources. Individual contributions have been identified by a numerical designation prefixed with the letter "c", after the name of the contributor, i. e. Lauren Jenkins (c2). A complete alphabetical listing of contributors, their titles, and affiliations are given in the Index of Contributors at the front of this volume, followed by c2, c6, etc to indicate the number(s) of their contribution(s). Figures reprinted from published sources are acknowledged in the figure captions. Numerical citations are used in order to keep the text less cluttered and easier to read, and are listed independently in the order cited at the end of each summary.

Chapter 6
Chromosome 1

Secondary constrictions in the human karyotype are less intensely stained by plain Giemsa than the rest of the chromosome. By G-banding, this region (designated the 1qh region, which typically corresponds to band q12) is more variably stained. It may appear as a Giemsa positive band or as a combination of darker and lighter bands (Fig. 6.1a). Long 1qh regions may appear as two or more dark bands separated by Giemsa-negative bands (Fig. 6.1a5). By C-banding, the entire region is usually Giemsa-positive (Figs. 6.1b and 6.2). Variations in the size of the secondary constriction of chromosome 1 by plain Giemsa were early noted to occur in 1/100 to 1/1000 newborns [1]. Variations in size by C-banding range from less than 1/2 the size of 16p (level 1,) to more than twice the size of 16p (level 5). From the New Haven study [2], 7.5% of children showed size variations by C-banding, 80% of which were *level 5* variants, and 20% of which were *level 1* variants (Fig. 3.1b).

Partial or complete inversions of 1qh (as seen by C-banding) were infrequent. In the New Haven study, partial inversions occurred with a frequency of 0.55% in white children and 0.07% in black children. Complete inversions of the 1qh region were not initially reported. However, Magenis et al. [3] reported on segregation of 1qh C-band variants from 42 families and classified them into 10 categories based on size and morphology (Fig. 6.2a). They found one complete inversion, sequentially stained by Q- and C-banding (Fig. 6.2b, c), for which they state, "… by C-banding alone might have been missed". Hsu et al. [4] subsequently reported 0.04–0.06% complete 1qh inversion in (American) black and Caucasian populations, and absence in Hispanic and Asian populations.

Additional examples of 1qh inversion are shown in Fig. 6.3. For the most part, no increases in fetal loss have been associated with 1qh inversions or with any of the other common 1qh heteromorphisms. However, one case that at first glance was not easily distinguishable from innocuous forms, has breakpoints in the euchromatin bands on either side of p and qh regions (Fig. 6.3f–h).

Heterogeneity is evident in 1qh heterochromatin not only by C-banding but also by Giemsa-11 staining (Fig. 6.1b) and by lateral asymmetry (Fig. 6.1c) [5]. Further 1qh heterogeneity is revealed by treatment with restriction enzymes (Alu1 and

H. E. Wyandt, V. S. Tonk, *Human Chromosome Variation: Heteromorphism and Polymorphism*, DOI 10.1007/978-94-007-0896-9_6,

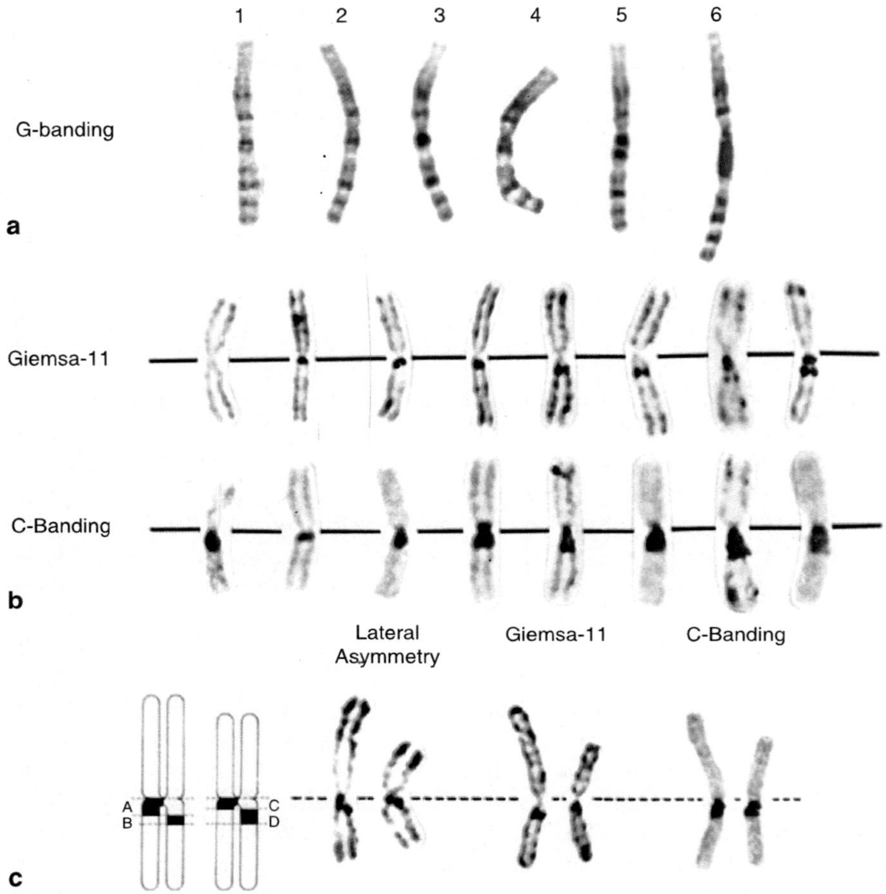

Fig. 6.1 **a** Normal no. 1's by *G-banding* showing increasing length of the secondary constriction, band 1q12. As the region increases in size, it appears to be divided into blocks, sometimes with a lighter proximal block (a2), a smaller dark-staining proximal block (a4), two clearly separated blocks of equal size and intensity (a5), or more rarely a small block and a very large block (a6). **b** *Giemsa-11* staining and *C-banding* variations from 8 individuals. *Top row*, from *left* to *right:* chromosome 1 with Giemsa-11-positive staining; five chromosomes with single Giemsa-11-positive bands at varying distances from the centromere; two chromosomes with two Giemsa-11-positive bands. *Bottom row:* C-banding of chromosome 1 from the same individuals. (Modified from Magenis et al. (1978), Science 202:64–65). **c** Further heterogeneity is revealed by *lateral asymmetry* in staining within the heterochromatin, due to interstrand compositional bias (i. e. T-rich strand vs. A-rich strand), revealed by substitution of T with 5-bromodeoxyuridine after 1 round of DNA synthesis. The longer chromosome on the left of each pair is a der(1)t(1;15) allowing it to be distinguished from the normal homolog (*right member of each pair*). Ideogram at the *left* depicts regions stained with the lateral asymmetry technique. *Blocks B and C* stain with Giemsa-11. *Blocks A and D* do not. [Modified from Magenis et al. (1978), Science 202:64–65. Contributors: a1–a3, Center for Human Genetics, Boston University (c1); *a4–a6*, Lauren Jenkins, Kaiser Permanente (c2)]

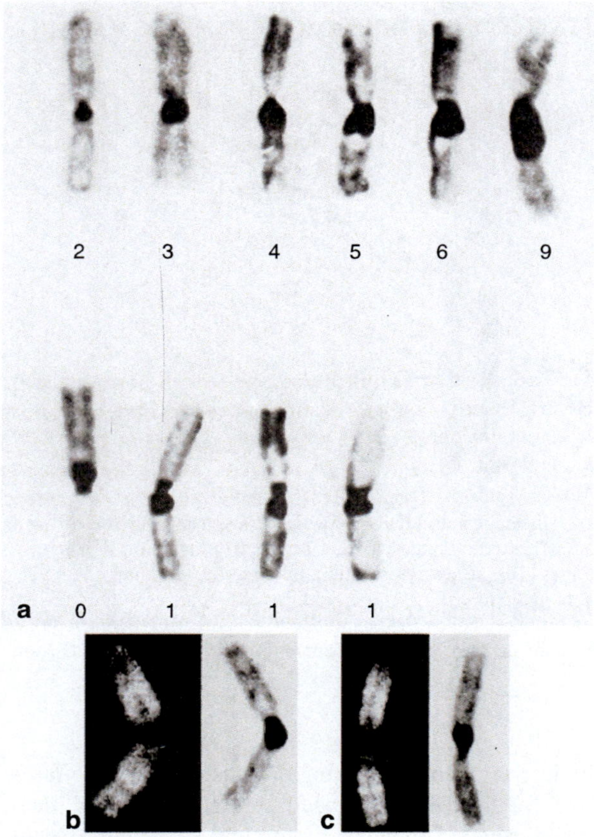

Fig. 6.2 **a** Variations of 1qh region by C-banding. Numbers (*top row*) are scores arbitrarily assigned by the investigators to designate size (*2–9*) and (*bottom row*) complete (*0*) or (*1*) partial inversion. (Reprinted with permission from Elsevier: from Magenis et al. (1977). In: Population Cytogenetics. Studies in Humans (EB Hook and IH Porter, eds.). Academic Press. New York. Figure 1, p 180). **b** and **c** Sequential staining of same chromosomes 1 by Q-banding (*left*) and C-banding (*right*) with typical 1qh region (*b*) and homolog with complete 1qh inversion (*c*) (Reprinted with permission from Elsevier: from Magenis et al. (1977). In: Population Cytogenetics. Studies in Humans (EB Hook and IH Porter, eds.). Academic Press. New York. Figure 4 (modified), p 183)

HaeIII) (Fig. 6.4). C-bands and Alu1-resistant bands correspond to each other, but bands remaining after *HaeIII* treatment are smaller, appear to exclude the proximal p11 and q11 regions, and correspond to DA/DAPI staining. G-11 stained regions appear to be sub-bands of DA/DAPI-positive regions [6, 7]. Heterogeneity in banding is also revealed by comparison of different fluorochromes such as DIPI and mithramycin (Fig. 6.5), which reportedly have different affinities for AT, or GC rich regions [8].

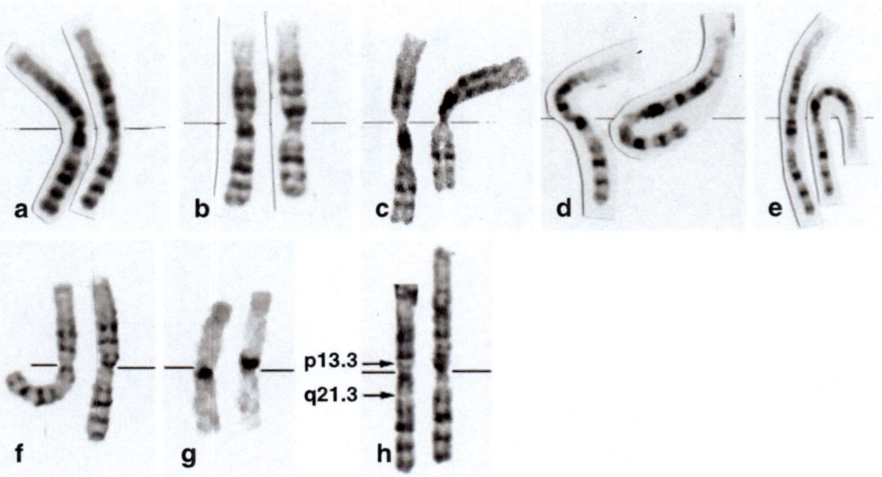

Fig. 6.3 **a–e** Pairs of 1's from four different individuals showing pericentric inversion of band q12 into the short arm. **f–h** Apparently unremarkable paternal pericentric inversion in a family with a history of multiple miscarriages (*f–g*) at lower resolution, by G- and C-banding, the inversion appears to be a typical complete inversion of the 1qh region; (*h*) at higher resolution, breakpoints were determined to be in euchromatic bands *p13.3* and *q21.3*, respectively. [Contributors: a, d, e Lauren Jenkins, Kaiser Permanente (c2); b, c, f–h Center for Human Genetics, Boston University (c1)]

Care should be taken in the interpretation of striking or unusual variants. Gardner et al. [9] cited six cases with 1qh+ (in non-banded chromosomes) and various anomalies, including two with severe anomalies resembling Patau and Meckel syndromes. In all six cases, both parents, one of which carried the same heteromorphism, were normal. Verma et al. [10, 11] reported two separate cases with speech delay and an unusual G-negative band inserted within the heterochromatic 1qh region that was positive by chromosome 1-specific painting, but parents were not studied in either case. As recently pointed out by Gardner and Sutherland [12], ascertainment bias is inevitable when performing chromosome studies based on clinical indications. Over-interpretation of the coincidental finding of a rare or unusual chromosome variant is not infrequent. As stated by these authors, "… if a normal parent or relative has the same unusual chromosome it should be seen as being harmless unless proven otherwise ….".

C-bands of chromosome 1 have been reported to be more variable in patients with malignant disease than in normal controls [13]. Rearrangements in the vicinity of the centromere are over-represented in many types of human cancer. Of interest is the characteristic rosette formation (Fig. 6.6) in a rare (non-cancer) disease

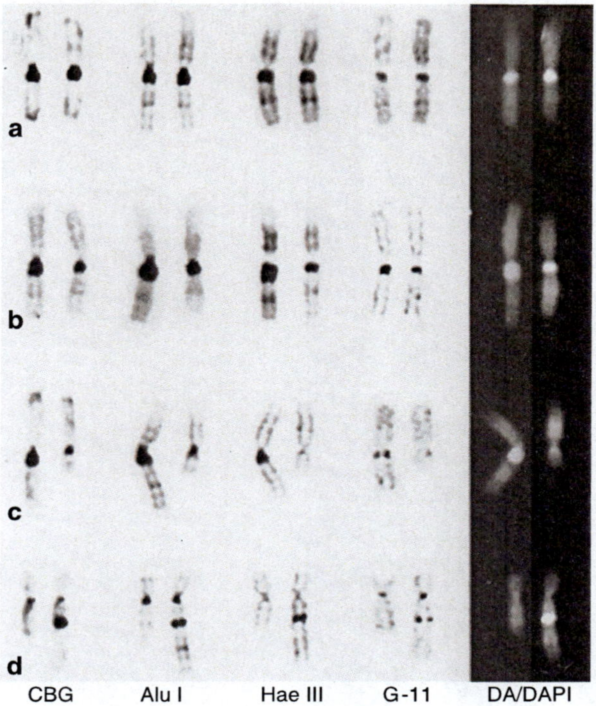

Fig. 6.4 The chromosomes stained for C-bands (*CBG*), for regions resistant to the restriction endonucleases *AluI* and *HaeIII* and by the *G-11* and *DA/DAPI* techniques. The rows *a* and *b* show the chromosomes from two normal subjects. The rows *c* and *d* are from (a balanced) translocation carrier (46,XY,t(1;7)(q11.1;q22)) showing the normal and derivative chromosomes 1 and 7, respectively. The derivative chromosomes are placed to the *right*. [Reproduced with permssion from Babu and Verma (1986), Histochem J 18:329–333]

called ICF (immune deficiency, centromeric instability and facial anomalies) [14], in which hypomethylation of DNA has been implicated [15, 16]. The alpha satellite region in chromosome 1 appears to be prone to breakage [17], and breaks in 1q12 heterochromatin show a lower than normal rate of DNA excision repair [18].

Euchromatic variants: In a review of euchromatic variants Jalal and Ketterling [19] refer to two case reports involving chromosome 1: one with a duplication of sub-bands 1q42.11–q42.12 in a short-statured child and an asymptomatic mother [20], and a second with a tandem duplication of bands p21–p31 in a fetus and normal mother (Fig. 6.7) [21]. In the latter case, the baby at 1, 2, and 3 months of age had normal physical examinations. As single case reports, each involves a different region, with neither completely fulfilling the criteria for a normal variant.

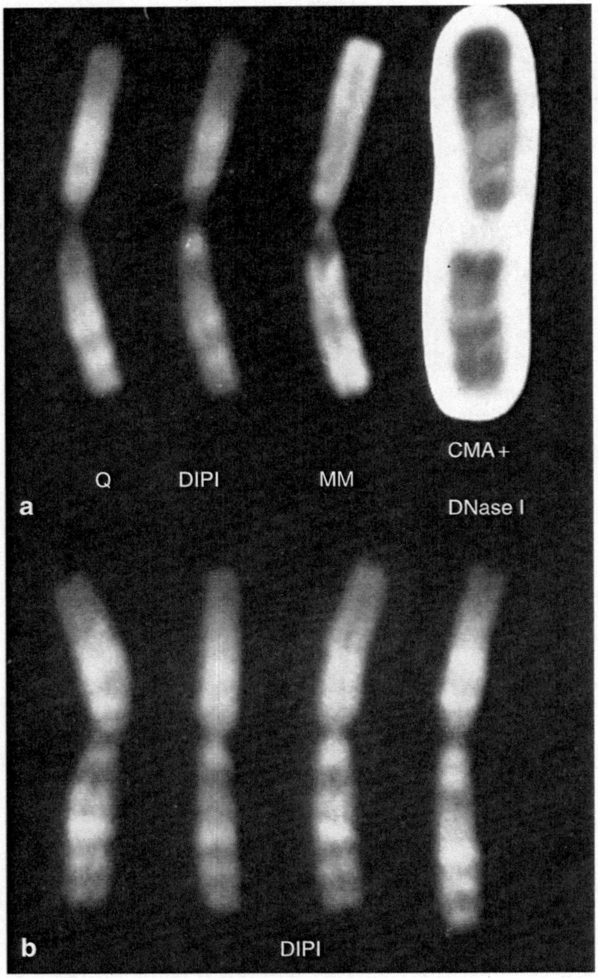

Fig. 6.5 **a** Three fluorescent images of the same chromosome 1 after subsequent quinacrine (*Q*), *DIPI* (*DIPI*=6-imidazolino-2(imidazolinophenyl) indole is closely related to *DAPI*, with similar staining affinity) and mithramycin (*MM*) treatment. Note that most of the material in the secondary constriction is *DIPI*-positive; a small segment close to the centromere is brightly stained by mithramycin and remains unstained by *DIPI*. On the right, another chromosome 1, treated with chromomycin A3 (*CMA*) and subsequently digested by *DNAse I* is shown; a reverse pattern is obtained by staining the remaining DNA with Giemsa or pinacyanol (for this preparation pinacyanol was used). **b** Chromosomes 1 from four different individuals, showing different lengths of the secondary constriction. The polymorphic material is *DIPI*-positive. [Reproduced with permission from Schnedl (1978). Hum Genet 41:1–9]

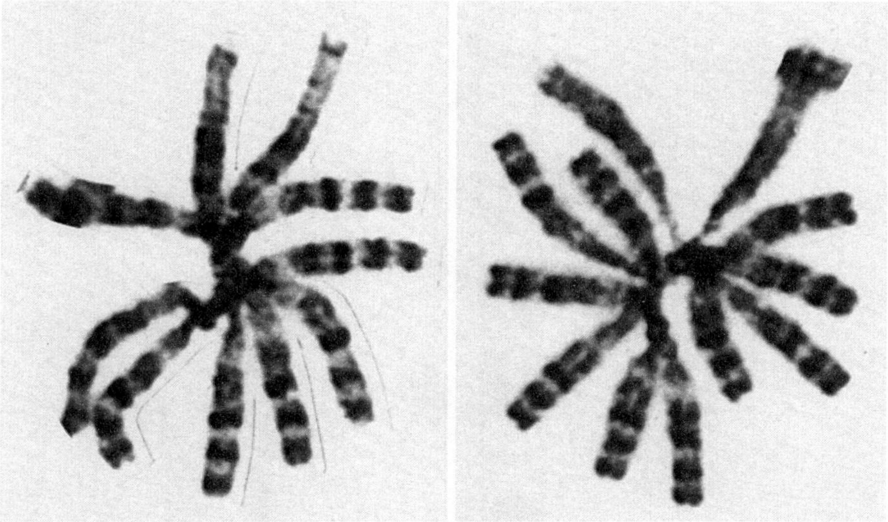

Fig. 6.6 Rosette formation of multiple copies of chromosome 1 attached at the 1qh region, associated with ICF syndrome. [Reproduced with permission from Sawyer et al. (1995). Am J Med Genet 56:203–209]

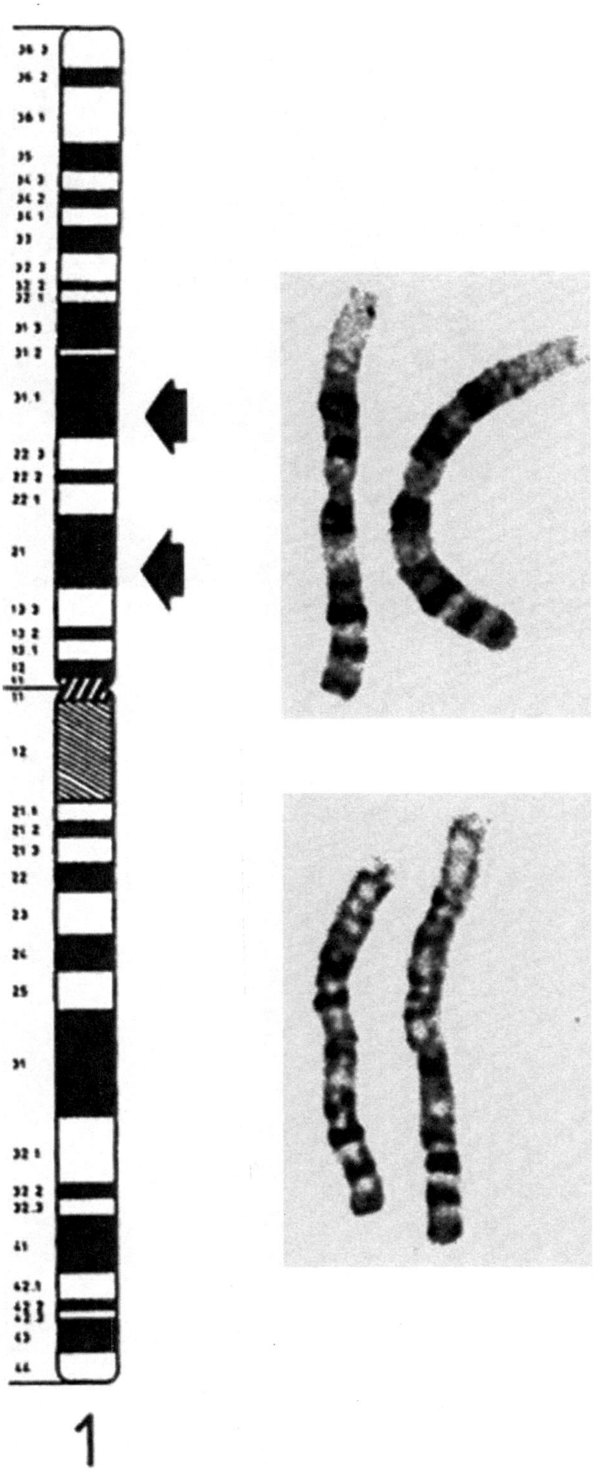

Fig. 6.7 Ideogram of chromosome 1 and trypsin-Giemsa banded partial chromosome numbers 1 from a fetus (*top*) and mother (*bottom*). [Reproduced with permission from Zazlov et al. (1993). Prenat Diagn 13:569–573]

References

1. Lubs HA, Ruddle FH (1971) Chromosome polymorphism in American negro and white populations. Nature 233:134–136
2. Lubs HA, Patil SR, Kimberling WJ, Brown J, Cohen M, Gerald P, Hecht F, Myrianthopoulos N, Summit RL (1977) Q and C-banding polymorphisms in 7 and 8 year old children: racial differences and clinical significance. In: Hook EB, Porter IH (eds) Population cytogenetic studies in humans. Academic Press, New York, pp 133–159
3. Magenis RE, Palmer CG, Wang L, Brown M, Chamberlin J, Parks M, Merritt AD, Rivas M, Yu PL (1977) Heritability of chromosome banding variants. In: Hook EB, Porter IH (eds) Population cytogenetics. Studies in humans. Academic Press, New York, pp 179–188
4. Hsu LYF, Benn PA, Tannenbaum HL, Perlis TE, Carlson AD (1987) Chromosomal polymorphisms of 1,9, 16 and Y in 4 major ethnic groups: a large prenatal study. Am J Med Genet 26:95–101
5. Magenis RE, Donlon TA, Wyandt HE (1978) Giemsa-11 staining of chromosome 1: a newly described heteromorphism. Science 202:64–65
6. Babu A, Verma RS (1986) Cytochemical heterogeneity of the C-band in human chromosome 1. Histochem J 18:329–333
7. Hedemann U, Schurmann M, Schwinger E (1988) The effect of restriction enzyme digestion of human metaphase chromosomes on C-band variants of chromosomes 1 and 9. Genome 30:652–655
8. Schnedl W (1978) Structure and variability of human chromosomes analyzed by recent techniques. Hum Genet 41:1–9
9. Gardner RJM, McCreanor HR, Paraslow MI, Veale AMO (1974) Are 1q+ chromosomes harmless? Clin Genet 6:383–393
10. Verma RS, Luke S, Brennan JP, Mathews T, Conte RA, Macera MJ (1993) Molecular topography of the secondary construction region (qh) of human chromosome 9 with an unusual euchromatic band. Am J Hum Genet 52:981–986
11. Verma RS, Kleyman SM, Conte RA (1997) An unusual G-negative band within 1qh region: a rare variant or an abnormality? Ann Genet 40:229–231
12. Gardner RJM, Sutherland, GR (2004) Chromosome abnormalities and genetic counseling, 3rd edn. Oxford University, Oxford
13. Wyandt HE, Patil RS (2004) Heteromorphisms in clinical populations. In: Wyandt HE, Tonk VS (eds) Atlas of human chromosome heteromorphisms. Kluwer, Dordrecht, pp 47–62
14. Sawyer JR, Swanson CM, Wheeler G, Cunniff C (1995) Chromosome instability in ICF syndrome: formation of micronuclei from multibranched chromomosomes 1 demonstrated by fluorescence *in situ* hybridization. Am J Med Genet 56:203–209
15. Jeanpierre M, Turleau C, Aurias A, Prieur M, Ledeist F, Fischer A, Viegas-Pequignot E (1993) An embryonic-like methylation pattern of classical satellite DNA is observed in ICF syndrome. Hum Mol Genet 2:731–735
16. Ji W, Hernandez R, Zhang XY, Qu GZ, Frady A, Varela M, Ehrlich M (1997) DNA demethylation and pericentromeric rearrangements of chromosome 1. Mut Res 379:33–41
17. Verma RS, Ramesh KH, Mathews T, Kleyman SM, Conte RA (1996) Centromeric alphoid sequences are breakage prone resulting in pericentromeric inversion heteromorphism of qh region of chromosome 1. Annal Genet 39(4):205–208
18. Surralles J, Darroudi F, Natarajan AT (1997) Low level DNA repair in human chromosome 1 heterochromatin. Genes Chromosomes Cancer 20:173–184
19. Jalal SM, Ketterling RP (2004). Euchromatic Variants. In: Wyandt HE, Tonk VS (eds) Atlas of human chromosome heteromorphisms. Kluwer, Dordrecht, pp 75–95
20. Bortotto L, Piovan E, Furlan R, Rivera H, Zuffardi O (1990) Chromosome imbalance, normal phenotype, and imprinting. J Med Genet 27:582–587
21. Zaslav AL, Blumenthal D, Fox JE, Thomson KA, Segraves R, Weinstein ME (1993) A rare inherited euchromatic heteromorphism on chromosome 1. Prenat Diagn 13:569–573

Chapter 7
Chromosome 2

Lubs et al. [1] found a detectably larger C-band-positive centromeric (cen+) region in chromosome 2 in 0.8% of 7- and 8-year-olds. No other studies have specifically addressed the frequency of 2cen+ in the normal population.

Pericentric inversion with breakpoints close to the centromere is a less frequent variant in chromosome 2. In initial non-banded studies of newborns, the frequency in the population was between 0.0001 and 0.013 [2, 3]. By banding, approximately 13 cases were found in 15,047 prenatal diagnoses [4, 5]. MacDonald and Cox [6] reviewed all banded cases from a six year period in a Canadian children's hospital and found two cases of inversion 2 among 3,619 patients referred for chromosome studies and an additional three cases among 1,820 prenatal chromosome studies. The overall frequency in prenatal studies is approximately 1 per 1,200. Most inv(2) s involve bands p11.2 in the short arm and q13 in the long arm (Fig. 7.1a–d) inv (a), both noted to involve common fragile sites [7]. Technically, because the band in the long arm of chromosome 2 that is inverted is non-heterochromatic, the inversion might be regarded as a structural abnormality. Djalali et al. [5] describe one case with breakpoints in p13 and q21 (Fig. 7.1e), and other exceptions have also been reported [8, 9]. Phillips [10] reported two families, one with the inv(2)(p11.2q13) and one with inv(2)(p13q11). In the first family there was no apparent effect on reproduction. In the second family some reproductive abnormalities were noted.

Djalali et al. [5] reviewed pedigrees of published cases of the common inversion and analyzed the data from cases referred for chromosome analysis and from prenatal diagnoses. From the pooled data, using Weinberg's proband method, they evaluated the risks for congenital anomalies and fetal wastage. In a corrected sample of 166 live offspring, born to carriers of pericentric inversions, 3.7% had heterogeneous phenotypic anomalies. Excluding two that died after delivery, this frequency was reduced to 3%, the same as the basic risk for abnormal offspring for any couple. Of 187 offspring, overall, 11.2% were fetal wastage (stillbirths and spontaneous miscarriages) and 85% had a normal phenotype. The rate of still-born and spontaneous abortions for carriers was about twice that of the general population. Male and female carriers occurred with equal frequency. They passed inv(2) on to about half of their children. About 3% of cases are de novo [11, 12]. There is no reported case of a recombinant chromosome resulting from inv(2)(p11.2q13) in a live-born child.

H. E. Wyandt, V. S. Tonk, *Human Chromosome Variation: Heteromorphism and Polymorphism*, DOI 10.1007/978-94-007-0896-9_7,

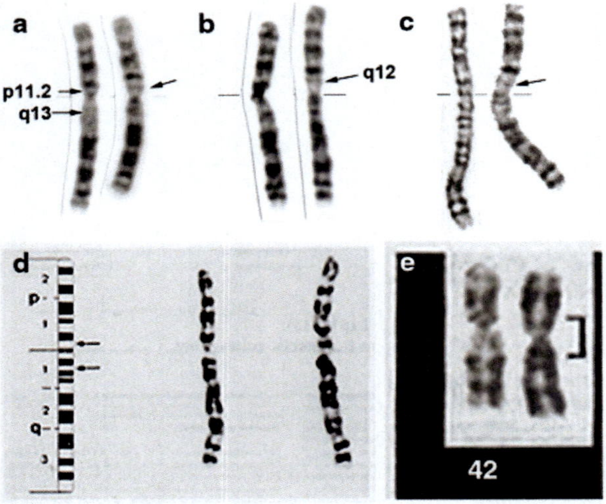

Fig 7.1 a–c Inversion in chromosome 2, inv(2)(p11.2q13) from three different cases. Normal homolog on left **a** shows breakpoints resulting in pericentric inversion (*right*). *Arrow* to the chromosome on the *right* in each pair points dark band 2q12 in the short arm of each pair (labeled in **b**). (Contributors: a, b Lauren Jenkins, Kaiser Permanente (c2); **c**, Center for Human Genetics, Boston University School of Medicine (c1). **d** Normal chromosome 2 (*left*) and inversion 2 (p11q13) (*right*). The ideogram with *arrows* shows the breakpoints. [Reproduced with permission from MacDonald IM, Cox DM (1985). Hum Genet 69:281–283]. **e** Example of inversion 2, Case 42, inv(2)(p13q21). [Reproduced in part with permission from Djalali M et al. (1986). Hum Genet 72:32–36]

This of course, in no way implies that inversions with more distal breakpoints are innocuous.

Although frequently found in normal individuals, inv(2) is often ascertained through individuals with mental retardation and/or congenital abnormalities [8]. However, none of the phenotypic abnormalities associated with inv(2) is consistent. A number of cases of inv(2) have also been associated with unrelated chromosome abnormalities [8, 9, 13–15]. Such associations presumably represent biases in ascertainment.

Euchromatic variants: A single case report of an interstitial deletion of 2q13-q14.1, detected by G-banding, was in a 38-year-old woman and her mother [16]. Both were phenotypically and clinically normal. As with other single case reports, further confirmation in additional cases is required to regard this as a normal variant. A different type of euchromatic deletion does appear to fulfill the criteria of a normal variant—namely, a number of cases of deletions of the subtelomeric region of 2q (not detectable cytogenetically) have been reported using subtelomeric FISH probes [17]. These cases, usually ascertained through an abnormal proband, typically reveal the same deletion in a parent. An indication that the deletion was clinically significant would be if one of the parents failed to have the same deletion.

References

1. Lubs HA, Patil SR, Kimberling WJ, Brown J, Cohen M, Gerald P, Hecht F, Myrianthopoulos N, Summit RL (1977) Q and C-banding polymorphisms in 7 and 8 year old children: racial differences and clinical significance. In: Hook EB, Porter IH (eds) Population cytogenetic studies in humans. Academic Press, New York, pp 133–159
2. Jacobs PA, Melville M, Ratcliffe S (1974) A cytogenetic survey of 11,680 newborn infants. Ann Hum Genet 37:359–376
3. Nielsen J, Sillesen I (1975) Incidence of chromosome aberrations in 11,148 newborn children. Hum Genet 30:1–12
4. Vejerslev LO, Friedrich U (1984) Experiences with unexpected structural chromosome aberrations in prenatal diagnosis in a Danish series. Prenat Diagn 4:181–186
5. Djalali M, Steinbach P, Bullerdiek J, Holmes-Siedel M, Verschraegen-Spae MR, Smith A (1986) The significance of pericentric inversions of chromosome 2. Hum Genet 72:32–36
6. MacDonald IM, Cox DM (1985) Inversions of chromosome 2 (p11p13): frequency and implications for genetic counseling. Hum Genet 69:281–283
7. Hecht F, Hecht BK (1984) Fragile sites and chromosome breakpoints in constitutional rearrangements. II. Spontaneous abortions, stillbirths and newborns. Clin Genet 26:174–177
8. Subrt I, Kozak J, Hnikova O (1973) Microdensitometric identification of the pericentric inversion of chromosome no. 2 and of duplication of the short arm of chromosome no. 7 in a reexamined case. Hum Hered 23:331–337
9. Hooft C, Coetsier H, Orye E (1968) Syndrome de Turner et inversion pericentrique probable du chromosome no. 2. Ann Genet (Paris) 11:181–183
10. Phillips RB (1978) Pericentric inversions inv(2)(p11q13) and inv(2)(p13q11) in two unrelated families. J Med Genet 15:388–390
11. Hesselbjerg U, Friedrich U (1979) Pericentric inversion in chromosome no. 2 as a de novo mutation. Humangenetik 53:117–119
12. Kozma C, Subasinghe C, Meck J (1996) Prenatal detection of de novo inversion of chromosome (2)(p13q11.2) and postnatal follow-up. Prenat Diagn 16:366–370
13. Cohen MM, Rosenmann A, Hacham-Zadeh S, Dahan S (1975) Dicentric X-isochromosome (Xqidic) and pericentric inversion of no. 2 inv(2)(p15q12) in a patient with gonadal dysgenesis. Clin Genet 8:11–17
14. Wikramanayake E, Renwick JH, Feguson-Smith MA (1971) Chromosomal heteromorphisms in the assignment of loci to particular autosomes: a study of four pedigrees. Ann Genet (Paris) 14:245–256
15. Verma RS, Dosik H, Wexler IB (1977) Inherited pericentric inversion of chromosome no. 2 with Robertsonian translocation (13q14q) resulting in trisomy for chromosome 13q. J de Genet Humaine 25:295–301
16. Sumption ND, Barber JCK (2001) A transmitted deletion of 2q13 to 2q14.1 causes no phenotypic abnormalities. J Med Genet 38:125–127
17. Ballif BC, Kashork CD, Shaffer L (2000) The promise and pitfalls of telomere region-specific probes. Am J Hum Genet 67:1356–1359

Chapter 8
Chromosome 3

Variation in size of the centromeric region of chromosome 3 (3cen) was initially described by C-banding [1, 2]. In the New Haven study of 7- and 8-year olds [1], C-band polymorphisms of 3cen were found in approximately 10% of cases (Fig. 8.1b). Q-banding revealed additional variations in size, location and intensity of the 3cen region. Q-bright centromeric regions (classified at intensity levels 4 or 5, Table 2.1) in the New Haven study were found in 57% of cases. Q-band heteromorphisms of chromosome 3 were studied in normal subjects for use in paternity testing [3], and revealed six different variants, including inv(3) (Fig. 8.1a).

Inv (3), a Q-bright, C-positive region in the short arm instead of the long arm of chromosome 3 (Fig. 8.1h) ranges in frequency from 0 to 11.1%, with an overall frequency of 1.26% [4–7]. The highest frequency (11.1%) was reported in a small selected retarded population from Estonia [4]. The highest frequency (8.3%) in a normal population was reported in a small selected Canadian population [5]. The investigators concluded that differences between normal and retarded populations were not significant.

Verma and Dosik [6] studied heteromorphisms of the centromeres of 3 and 4 in 100 Caucasians by sequential QFQ (Q-banding) and RFA banding (reverse banding by acridine orange). QFQ banding showed 62% of 3cen and 15% of 4cen with intensities ≥ level 3. Color variants by RFA could not be distinguished in a blind study. In studies of American blacks [7–9], the frequency of QFQ-bright 3cen (levels 3 and 4) was 54.5%. Again, neither RFA differences nor C-banding differences could be discerned in a blind study.

Euchromatic variants: A terminal deletion with a breakpoint at 3p25.3 was reported in a fetus and a phenotypically normal mother [10]. At term, a normal baby girl was born. Although of interest, we can only reiterate our previous precautions regarding single case reports.

H. E. Wyandt, V. S. Tonk, *Human Chromosome Variation: Heteromorphism and Polymorphism,* DOI 10.1007/978-94-007-0896-9_8,

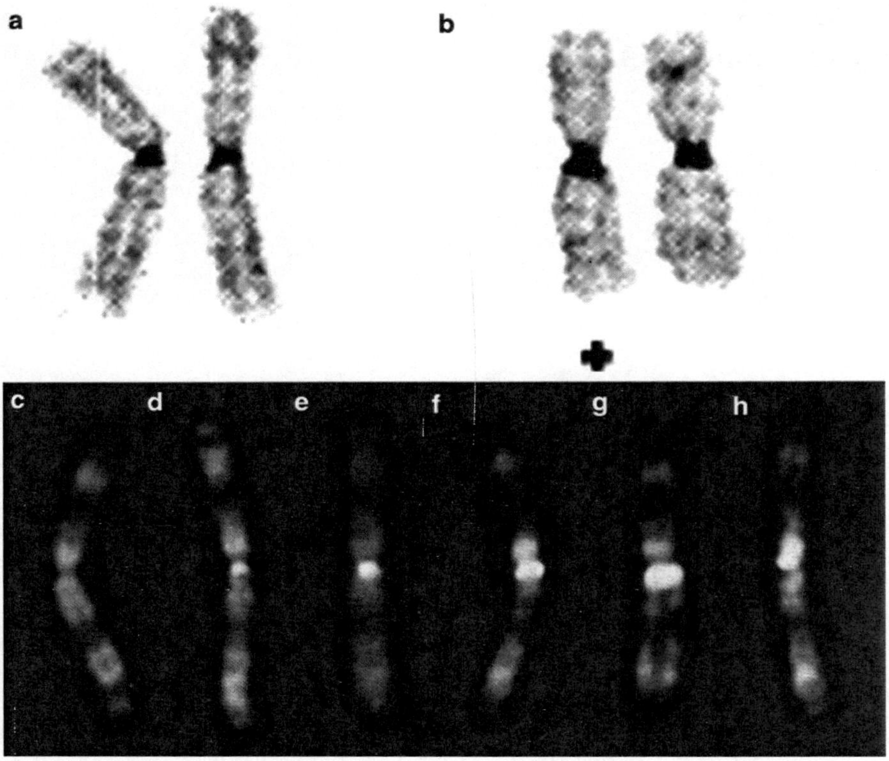

Fig. 8.1 **a** C-banding of 3cen in normal pair of homologs; **b** C-banding of 3cen showing slightly larger C-band variant in one homolog than the other. [Reproduced from Craig-Holms, Moore, Shaw (1973). Polymorphism of human C-band heterochromatin I: frequency of variants. Am J Hum Genet 25:181–192]. **c–h** Variation in intensity, size and location of 3cen variants by Q-banding, including apparent inversion of a Q-bright variant (h) into the short arm. [Reproduced from Olson et al. (1986). Am J Hum Genet 38:235–252]

References

1. Lubs HA, Patil SR, Kimberling WJ, Brown J, Cohen M, Gerald P, Hecht F, Myrianthopoulos N, Summit RL (1977) Q and C-banding polymorphisms in 7 and 8 year old children: racial differences and clinical significance: In: Hook EB, Porter IH (eds) Population cytogenetic studies in humans. Academic Press, New York, pp 133–159
2. McKenzie WH, Lubs HA (1975) Human Q and C chromosomal variations: distribution and incidence. Cytogenet Cell Genet 14:97–115
3. Olson SB, Magenis RE, Lovrien EW (1986) Human chromosome variation: the discriminatory power of Q-band heteromorphism (variant) analysis in distinguishing between individuals, with specific application to cases of questionable paternity. Am J Hum Genet 38:235–252
4. Mikelsaar AV, Kaosaar ME, Tuur SJ, Viikmaa MH, Talvik TA, Laats J (1975) Human karyotype polymorphisms. III. Routine and fluorescence microscope investigation of chromosomes in normal adults and mentally retarded children. Humangenetik 26:1–23

5. Fogle TA, McKenzie WH (1980) Cytogenetic study of a large Black kindred: inversions, heteromorphisms and segregation analysis. Hum Genet 58:345–352
6. Verma RS, Dosik H (1979) Frequencies of centromeric heteromorphisms of human chromosomes 3 and 4 as detected by QFQ technique: can they be identified by RFA technique? Can J Genet Cytol 21:109–113
7. Verma RS, Dosik H (1980) Human chromosomal heteromorphisms in American blacks. I. Structural variability of chromosome 3. J Hered 71:441–442
8. Verma RS, Dosik H (1981) Human chromosomal heteromorphisms in American blacks. IV. Intensity variation in centromeric regions of chromosomes 3 and 4. Can J Cytol 23:315–320
9. Conte RA, Luke S, Verma RS (1992) Molecular characterization of "inverted" pericentromeric heterochromatin of chromosome 3. Histochemistry 97:509–510
10. Knight LA, Yong MH, Tan M, Ng ISL (1995) Del(3)(p25.3) without phenotypic effect. J Med Genet 32:994–995

Chapter 9
Chromosome 4

In the study of 7- and 8-year olds [1], Q-band heteromorphisms consisting of level 4 and 5 intensities (Table 2.1) were seen in about 20% of the white children and 9% of the black children. C-band variants of 4cen were seen in approximately 7.6% of children overall with no significant racial difference. Verma and Dosik [2] found bright QFQ variants (levels 3 and 4) in 7% of American blacks. Bardham et al. [3] found two types of common QFQ-band variants, one showing an intensely Q-bright band in 4cen and the other an intensely Q-bright band in proximal 4p. A higher frequency of Q-bright heteromorphisms reportedly found in patients with a variety of clinical problems has not been supported in subsequent studies [4, 5]. In a study of Q-band heteromorphisms for paternity testing [6], seven different variants were found in normal subjects (Fig. 9.1a).

In addition to the Q heteromorphisms described above, a rare variant involving an unusually large G-band-negative and C-band-positive block of centromeric heterochromatin (Fig. 9.1b) was described by Docherty and Bowser-Riley [4]. McKenzie and Lubs [7] described cases with a large C-positive region on 4, in which the paracentromeric region was Q-bright, but 4cen was non-fluorescent. Hence, the variant would be missed if C-banding was not done. Babu and Verma [5] studied a small Q-bright paracentromeric band in 4p, which manifested as a large CBG-positive band that included both centromeric and paracentromeric heterochromatin. When the same chromosome was stained with Giemsa after AluI digestion, only the paracentromeric region was Giemsa-positive.

By molecular studies, two alpha satellite DNAs have been cloned and analyzed for chromosome 4 [8]. Under stringent wash conditions, they hybridized only to the pericentromeric region of chromosome 4; under non-stringent conditions they hybridized to an alpha satellite supra-family that included chromosomes 2, 4, 8, 9, 13, 14, 15, 18, 20, 21 and 22. Southern blot yielded a single 3.2 kb higher order MspI restriction fragment or a combination of a 2.6 and 0.6 kb MspI fragments, which together constituted about 60% of the length of the larger fragment.

Euchromatic Variants: Two unusual cases involving euchromatic bands are shown. In one case, a phenotypically normal individual has an apparent paracentric inversion in 4p16 with breaks in 4p16.1 and 4p16.3 (Fig. 9.1c). By FISH analysis (not shown), a probe for Wolf-Hirschhorn region was split by the inversion. A second

H. E. Wyandt, V. S. Tonk, *Human Chromosome Variation: Heteromorphism and Polymorphism*, DOI 10.1007/978-94-007-0896-9_9,

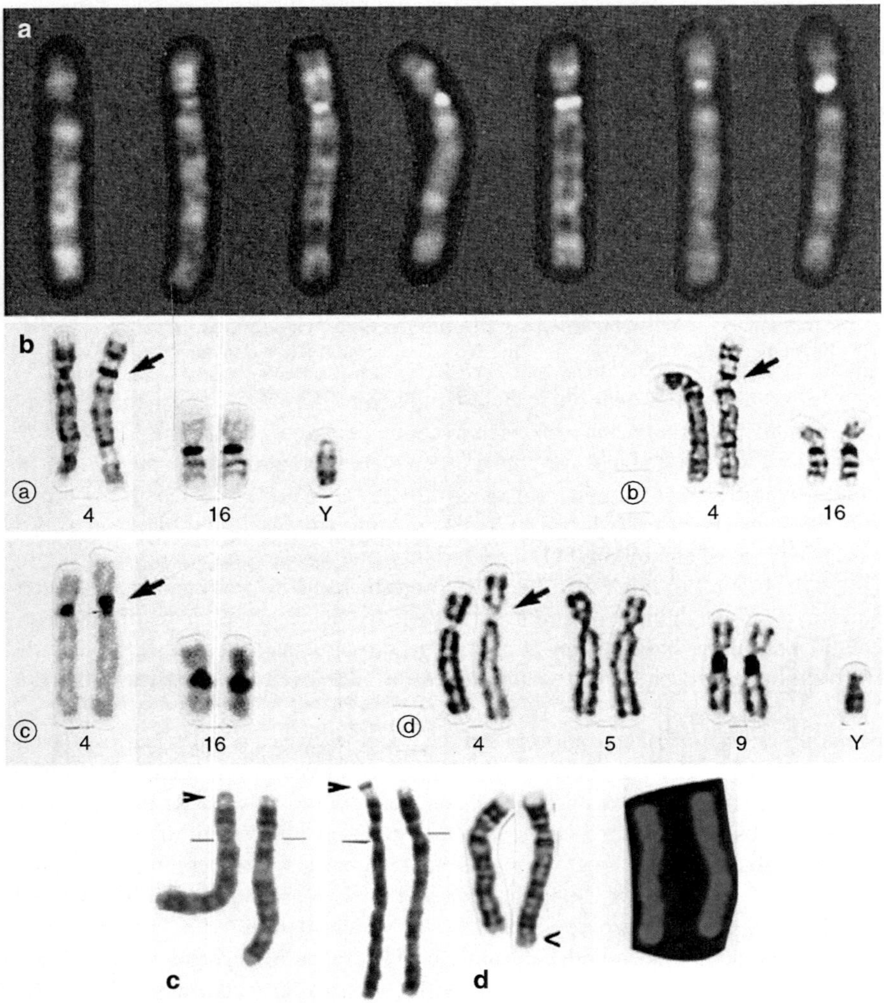

Fig. 9.1 **a** Common Q-band heteromorphisms of 4cen. (Figure 2 from Olson et al. (1986). Am J Hum Genet 38:235–252). **b** Partial karyotypes of antenatally screened fetus (*a, d*) and of the father (*b, c*) showing the variant chromosome 4 (*arrows*). (*a*) G-banding using trypsin after pretreatment with acetic saline. (*b*) G-banding using trypsin after treatment with hydrogen peroxide. (*c*) C banding. (*d*) Silver staining. (Reproduced with permission from Docherty, Bowser-Riley (1984). J Med Genet 21:470–472). **c, d** Euchromatic variants: (*c*) Rare variant in chromosome 4, inv(4)(p16.1p16.3). [Contributor: Center for Human Genetics, Boston University (c1)]; (*d*) Rare variant in chromosome 4, dup(4)(q35q35). [Contributor: Steve Gerson, Dianon Systems, Inc. (c18)]

case (Fig. 9.1d) has additional material at 4q35 in a fetus and phenotypically normal father without other apparent phenotypic abnormalities. Information in both of these instances is provided with the usual precaution regarding single case reports.

References

1. Lubs HA, Patil SR, Kimberling WJ, Brown J, Cohen M, Gerald P, Hecht F, Myrianthopoulos N, Summit RL (1977) Q and C-banding polymorphisms in 7 and 8 year old children: racial differences and clinical significance. In: Hook EB, Porter IH (eds) Population cytogenetic studies in humans. Academic Press, New York, pp 133–159
2. Verma RS, Dosik H (1981) Human chromosomal heteromorphisms in American blacks. IV. Intensity variation in centromeric regions of chromosomes 3 and 4. Can J Cytol 23:315–320
3. Bardham S, Singh DN, Davis K (1981) Polymorphism in chromosome 4. Clin Genet 20:44–47
4. Docherty Z, Bowser-Riley SM (1984) A rare heterochromatic variant of chromosome 4. J Med Genet 21:470–472
5. Babu A, Verma RS (1986) Heteromorphic variants of human chromosome 4. Cytogenet Cell Genet 41:60–61
6. Olson SB, Magenis RE, Lovrien EW (1986) Human chromosome variation: the discriminatory power of Q-band heteromorphism (variant) analysis in distinguishing between individuals, with specific application to cases of questionable paternity. Am J Hum Genet 38:235–252
7. McKenzie WH, Lubs HA (1975) Human Q and C chromosomal variations: distribution and incidence. Cytogenet Cell Genet 14:97–115
8. Mashkova TD, Akopian TA, Romanova LY, Mitkevich SP, Yurov SP, Kisselev LL, Alexandrov IA (1994) Genomic organization, sequence and polymorphism of the human chromosome 4-specific alpha satellite DNA. Gene 140:211–217

Chapter 10
Chromosome 5

A rare variant of chromosome 5, detected prenatally, consisted of a large pericentromeric block of heterochromatin that was C and G-11 positive. It was also present in several family members, all of whom were phenotypically normal (Fig. 10.1) [1]. In a more recent report of a three-generation family [2], FISH studies showed a similar variant that hybridized with satellite III sequences but not with beta satellite sequences of chromosome 9. Earlier studies [3] had shown G-11 staining to correlate to the locations of satellite III on several chromosomes, especially on chromosome 9.

Molecular analysis of centromeric and pericentromeric sequences of chromosome 5 have shown alpha satellite DNA's D5Z1 and D19Z2 that hybridize to chromosomes 5 and 19 centromeric regions, respectively, to have high sequence homology that can only be separated by pulse field gel electrophoresis [4]. Hence, cross-hybridization by FISH for alpha satellite sequences occurs for chromosomes 5 and 19. Cross-hybridization also occurs for a different subset of alphoid sequences for chromosomes 1 and 19 [5].

Euchromatic Variants: A duplication of 5q15-q21 has been described in a phenotypically normal father and in monozygotic twin daughters with different abnormal phenotypes [6]. The breakpoints were determined to be the same in the father and the twins with cosmid probes spanning the 5q13-23 region. The anomalies of the twin daughters were considered to be unrelated to the familial duplication.

In their review, Jalal and Ketterling [7] describe several reports [8–11] of 5p interstitial deletions of different sizes with phenotypes ranging from normal to mental retardation and inconsistent abnormalities. From combined observations, loss of part or all of 5p14 appears to have no adverse phenotypic effect. However, if the deletion extends into the band 5p13 or band 5p15, dysmorphism including mental retardation results (Fig. 10.2).

H. E. Wyandt, V. S. Tonk, *Human Chromosome Variation: Heteromorphism and Polymorphism,* DOI 10.1007/978-94-007-0896-9_10,

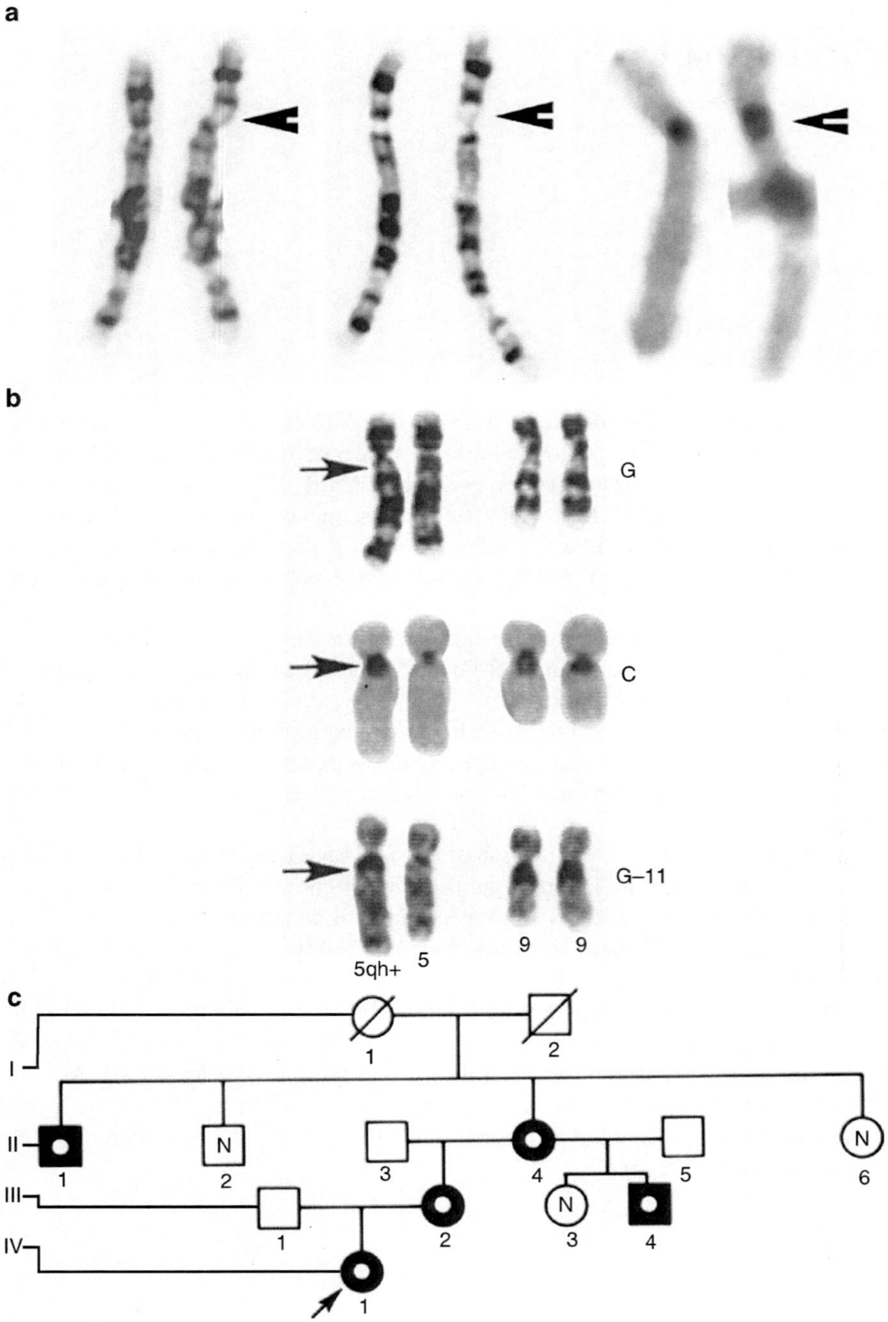

Fig. 10.1 **a** Large 5cen by G-banding (*arrow*) compared with normal homolog (*left*) in two pairs of chromosome 5s (*left and middle*) and chromosomes 5 pair from same patient by C-banding (*far right; arrow* points to larger centromeric region). [Contributor: Lauren Jenkins, Kaiser Permanente (c2)]. (**b**) Pairs of chromosome 5 (one with large 5qh region) and 9 by G, C and G-11 banding. (**c**) Pedigree of family. (Figures. 1 and 2 reproduced with permission from Fineman RM et al. (1989). Am J Med Genet 32:498–499)

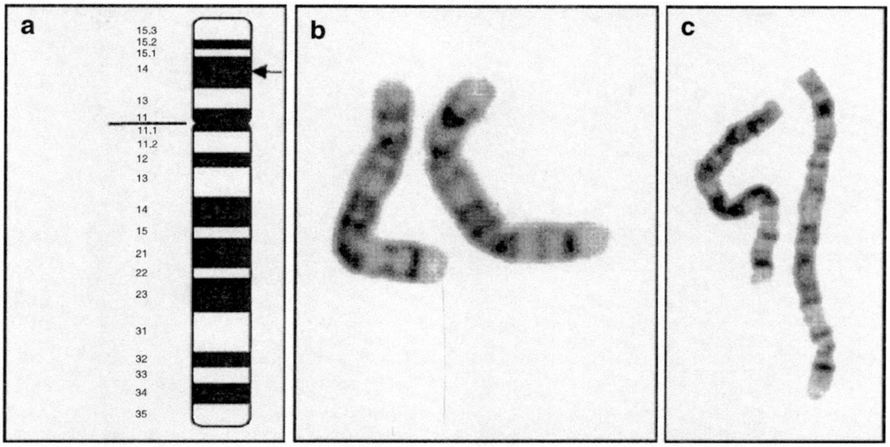

Fig. 10.2 Euchromatic variant of chromosome 5 showing apparent deletion of 5p14 in child and mother. [Reproduced with permission from Hand JL et al. (2000). Prenat Diagn 20:144–148]

References

1. Fineman RM, Issa B, Weinblatt V (1989) Prenatal diagnosis of a large heterochromatic region in a chromosome 5: implications for genetic counseling. Am J Med Genet 32:498–499
2. Doneda L, Gandolfi P, Nocera G, Larizza L (1998) A rare chromosome 5 heterochromatic variant derived from insertion of 9qh satellite 3 sequences. Chromosome Res 6:411–414
3. Buhler EM, Tsuchimoto T, Jurik LP, Stalder GR (1975) Satellite DNA III and alkaline Giemsa staining. Humangenetik 26:329–333
4. Peuchberty J, Laurent A-M, Mimenez S, Billault A, Brun-Laurent M-E, Calenda A, Marcais B, Prades C, Ioannou P, Yurov Y, Roizès G (1999) Genetic and physical analyses of the centromeric and pericentromeric regions of human chromosome 5: recombination across 5cen. Genomics 56:274–287
5. Finelli P, Antonacci R, Marzella R, Lonoce A, Archidiacono N, Rocchi M (1996) Structural organization of multiple alphoid subsets coexisting on human chromosomes 1, 4, 5, 9, 15 and 19. Genomics 38:325–330
6. Shuan-Yow L, Gilbson LH, Gomez K, Pober BR, Yang-Feng TL (1998) Familial dup(5)(q15q21) associated with normal and abnormal phenotypes. Am J Med Genet 75:75–77
7. Jalal SM, Ketterling RP (2004) Euchromatic variants. In: Wyandt HE and Tonk VS (eds) Atlas of human chromosome heteromorphisms. Kluwer, Dordrecht, pp 75–95
8. Walker JL, Blank CE, Smith BAM (1984) Interstitial deletion of the short arm of chromosome 5 in a mother and three children. J Med Gen 21(6):465–467
9. Overhauser J, Golbus MS, Schonberg SA, Wasmuth JJ (1986) Molecular analysis of an unbalanced deletion of the short arm of chromosome 5 that produces no phenotype. Am J Hum Genet 39:1–10
10. Keppen LD, Gollin SM, Edwards D, Sawyer J, Wilson W, Overhauser J (1992) Clinical phenotype and molecular analysis of a three-generation family with an interstitial deletion of the short arm of chromosome 5. Am J Med Genet 44:356–360
11. Hand JL, Michels VV, Marinello MJ, Ketterling RP, Jalal SM (2000) Inherited interstitial deletion of chromosomes 5p and 16q without apparent phenotypic effect: further confirmation. Prenat Diagn 20:144–148

Chapter 11
Chromosome 6

Heteromorphism of the centromeric region of chromosome 6, as seen by G- and C-banding (Fig. 11.1a), was found in the New Haven study in about 3% of 7- and 8- year-olds [1]. About half had a small centromeric region and half a larger than average centromeric region. In a study of 92 consecutive individuals referred for cytogenetic testing, 6p11 was enlarged in size in approximately 9% of subjects [2]. C-band heteromorphism of chromosome 6 in several families was used to establish linkage of the HLA region to chromosome 6 [3].

Alpha satellite sequences in the centromeric region of chromosome 6 have been isolated and characterized [4]. In two families a C-band-positive 6ph+ region (Fig. 11.1b) was two to three times the size of the region in the homolog. In the first family [5], a tritium labeled alphoid repeat, D6Z1 (p308) of 6ph+ was amplified 2-3-fold by *in situ* hybridization and by dot blot hybridization. A similar 6q11+ variant detected prenatally in a fetus and in the mother [6] was G-, Q- and DA/DAPI-negative, but C-band-positive. *In situ* hybridization revealed alpha satellite D6Z1 was amplified. The region was late replicating and revealed lateral asymmetry after one cycle of 5-BrdU. The region in neither family was affected by 5-azacytidine or distamycin.

Euchromatic variants: Jalal and Ketterling [7] review two interesting cases with euchromatic deletions in chromosome 6. One, a de novo interstitial deletion of 6q23.3–6q24.2, is in a developmentally normal 3-year-old girl with mild dysmorphic features [8]. The second, a 2-year-old girl with an interstitial deletion of 6q23.1-q24.2 and growth retardation (3rd centile), has dysmorphic features unlike those in the first case [9]. According to Jalal and Ketterling, the latter authors reportedly reflect that the differences in phenotype may be due either to haploinsufficiency of genes within 6q23.3–6q24.2 or to imprinting of paternal genes. Chromosome 6 imprinting has been implicated in neonatal diabetes mellitus [10]. Confirmation that deletion of 6q23.1–6q24.2 is a normal variant awaits further reports.

H. E. Wyandt, V. S. Tonk, *Human Chromosome Variation: Heteromorphism and Polymorphism,* DOI 10.1007/978-94-007-0896-9_11,

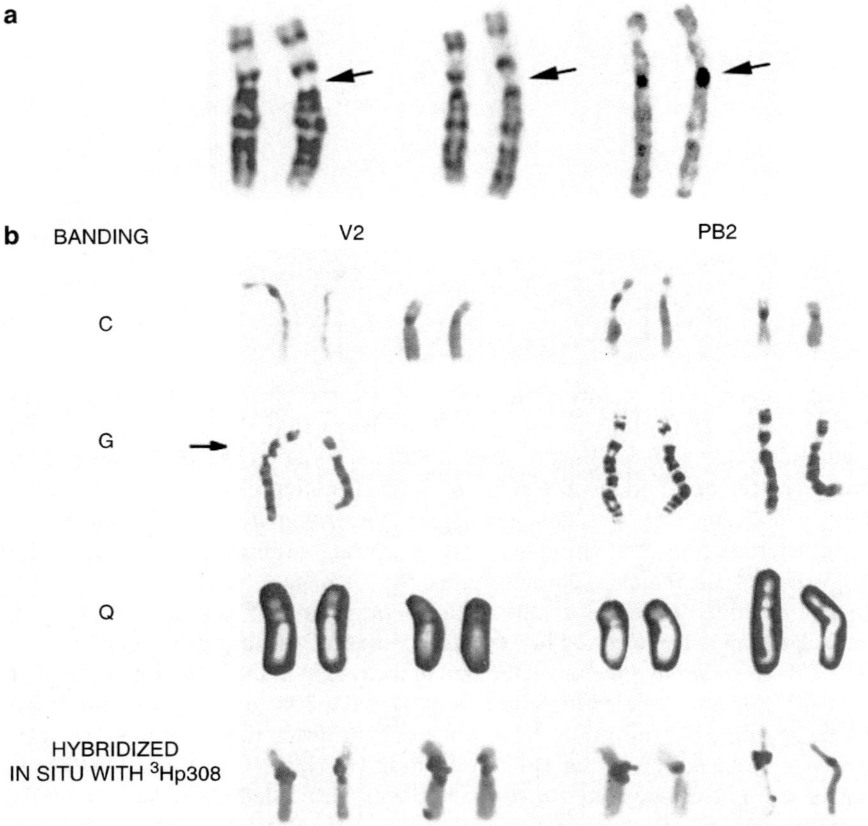

Fig. 11.1 **a** Pairs of chromosome 6 (*left and middle*) showing large centromere (right member of each pair) by G-banding. Pair of chromosome 6 (*right*) by C-banding from same patient. [Contributor: Lauren Jenkins, Kaiser Permanente (c2)]. **b** Pairs of chromosome 6, from two different families (V2 and PB2) showing large 6cen (left member of each pair) by C, G and Q-banding. Last row: hybridization shows amplification of the tritiated alpha satellite sequence, ^{3}Hp308. [Reproduced with permission from Jabs EW and Carpenter N (1988). Am J Hum Genet 43:69–74]

References

1. Lubs HA, Patil SR, Kimberling WJ, Brown J, Cohen M, Gerald P, Hecht F, Myrianthopoulos N, Summit RL (1977) Q and C-banding polymorphisms in 7 and 8 year old children: racial differences and clinical significance. In: Hook EB, Porter IH (eds) Population cytogenetic studies in humans. Academic Press, New York, pp 133–159
2. Madan K, Bruinsman AH (1979) C-band polymorphism in human chromosome no. 6. Clin Genet 15:193–197
3. Polacek LA, Phillips RB, Hackbarth SA, Duquesnoy RJ (1983) A linkage study of the HLA region using C-band heteromorphisms. Clin Genet 23:177–185

4. Jabs EW, Wolf SF, Migeon BR (1984) Characterization of a cloned DNA sequence that is present at the centromeres of all human autosomes and the X chromosome and shows polymorphic variation. Proc Natl Acad Sci U S A 81:4884–4888
5. Jabs EW, Carpenter N (1988) Molecular cytogenetic evidence for amplification of chromosome-specific alphoid sequences at enlarged C-bands on chromosome 6. Am J Hum Genet 43:69–74
6. Lin MS, Zhang A, Fujimoto A, Wilson MG (1994) A rare 6q11+ heteromorphism: cytogenetic analysis and *in situ* hybridization. Hum Hered 44:31–36
7. Jalal SM, Ketterling RP (2004) Euchromatic Variants. In Wyandt HE, Tonk VS (eds) Atlas of human chromosome heteromorphisms. Kluwer Academic, Dordrecht, pp 75–95
8. Kumar A, Cassidy SB, Romero L, Schwartz S (1999) Molecular cytogenetics of a de novo interstitial deletion of chromosome arm 6q in a developmentally normal girl. Am J Med Genet 86:227–231
9. Matkins SV, Meyer JE, Berry AC (1987) A child with partial monosomy 6q secondary to a maternal direct insertional event. J Med Genet 24:227–229
10. Christian SL, Rich BH, Loebl C, Israel J, Vasa R, Kittikamron K, Spiro R, Rosenfield R, Ledbetter DH (1999) Significance of genetic testing for paternal uniparental disomy of chromosome 6 in neonatal diabetes mellitus. J Pediatr 134:42–46

Chapter 12
Chromosome 7

An enlarged centromeric region, reported in about 4% of individuals [1], is generally of no known clinical significance. Of interest, is one reported case (Fig. 12.1) of uniparental disomy for a chromosome 7 with a cen+ heteromorphism that was maternal in origin [2]. The 4-year-old son had cystic fibrosis (CF) and very short stature. DNA polymorphisms of paternal loci spanning chromosome 7 were negative in the child. Uniparental disomy theoretically occurs in the population with a frequency of 1/30,000 [3, 4]. However, the frequency in CF may be considerably higher. Two reported cases of maternal uniparental disomy with CF suggest their origin was due to a nullisomic sperm with maternal isodisomic rescue [2].

Molecular analyses indicate the alpha satellite DNA on chromosome 7 consists of two distinct alphoid domains, arranged in higher order monomers [5, 6]; D7Z2 on the short arm side and D7Z1on long arm side. In a screen for centromeric sequences, four clones were found that contained alpha satellite, L1 repeat units, an Alu element and a novel AT-rich repeated sequence [7, 8].

H. E. Wyandt, V. S. Tonk, *Human Chromosome Variation: Heteromorphism and Polymorphism,* DOI 10.1007/978-94-007-0896-9_12,

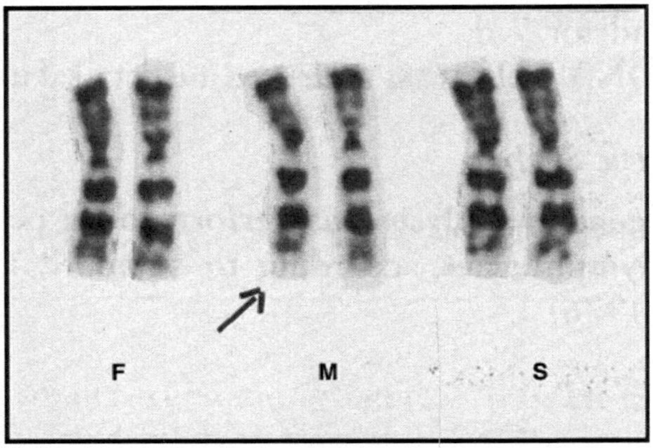

Fig. 12.1 Heteromorphism in a mother and in her son who is homozygous for 7cen+. Molecular studies indicated the son has uniparental disomy for the maternal chromosome 7. [Reproduced with permission from Voss et al., Am J Hum Genet 45:373–380 (1989)]

References

1. Lubs HA, Patil SR, Kimberling WJ, Brown J, Cohen M, Gerald P, Hecht F, Myrianthopoulos N, Summit RL (1977) Q and C-banding polymorphisms in 7 and 8 year old children: racial differences and clinical significance. In: Hook EB, Porter IH (eds) Population cytogenetic studies in humans. Academic, New York, pp 133–159
2. Voss R, Ben-Simon E, Avital A, Godfrey S, Zlotogora J, Dagan J, Tikochinski Y, Hillel J (1989) Isodisomy of chromosome 7 in a patient with cystic fibrosis: could uniparental disomy be common in humans? Am J Hum Genet 45:373–380.
3. Warburton D (1988) Editorial: uniparental disomy; a rare consequence of the high rate of aneuploidy in human gametes. Am J Hum Genet 42:215–216
4. Spence JE, Periaccante RG, Greig GM, Willard HF Ledbetter DH, Hejmancik JF, Pollack MS et al (1988) Uniparental disomy as a mechanism for human genetic disease. Am J Hum Genet 42:217–226
5. Waye JS, England SB, Willard HF (1987) Genomic organization of alpha satellite DNA on human chromosome 7: evidence for two distinct alphoid domains on a single chromosome. Mol Cell Biol 7:349–356
6. Fetni R, Richer CL, Malfoy B, Dutrillaux B, Lemieux N (1997) Cytological characterization of two distinct alpha satellite DNA domains on human chromosome 7, using double-labeling hybridizations in fluorescence and electron microscopy on a melanoma cell line. Cancer Genet Cytogenet 96:17–22
7. Wevrick R, Willard HF (1991) Physical map of the centromeric region of human chromosome 7: relationship between two distinct alpha satellite arrays. Nucleic Acids 19:2295–2301
8. Wevrick R, Willard VP, Willard HF (1992) Structure of DNA near long tandem arrays of alpha satellite DNA at the centromere of human chromosome 7. Genomics 14:912–923

Chapter 13
Chromosome 8

Centromeric variants of chromosome 8 appear to be rare. In 7- and 8-year-olds [1], only one case had a large chromosome 8 centromere (Fig. 13.1a). An earlier study [2] also reported one case with a C-band-positive 8cen+.

Euchromatic variants: Euchromatic variants involving the distal band of 8p, have been reported in more than a dozen families [3]. Barber et al. [4] described duplication of band 8p23.1 in 18 individuals from seven families, with the duplication transmitted from parents to children in four of the families. Duplications were confirmed by FISH with a YAC clone that mapped to 8p23.1. A review of reported cases determined that 25 of 27 duplication carriers, including eight cases analyzed in the prenatal/neonatal time period, had no evidence of phenotypic anomalies. The probands were dysmorphic in two families. The authors concluded these duplications do not have clinical significance (Fig. 13.1b). A duplication of 8p23.1p23.3 is reported in a 34 year-old male with oligoasthenozoospermia. His mother and a 27 year-old brother, who also had the duplication, were phenotypically normal [5].

Deletions in the region 8q24.11–8q24.13 are associated with Langer-Giedion (LGS) or trichorhinophalangeal syndromes, but deletions in 8q24 have been reported with variable phenotypes ranging from lethal to normal [3]. One case with LGS features, but normal intelligence, had an interstitial deletion of 8q24.11–8q24.12 [6]. An interstitial deletion of 8q24.13–8q24.22 was reported in a phenotypically normal mother and in her stillborn fetus. Fetal abnormalities were not evident by prenatal ultrasound [7], but there was a history of pregnancy loss in other family members. FISH revealed deletion of c-myc, but no family history of hematological disorder. Two de novo cases of interstitial deletions in distal 8q24.1 had mild anomalies, but not LGS [8].

H. E. Wyandt, V. S. Tonk, *Human Chromosome Variation: Heteromorphism and Polymorphism*, DOI 10.1007/978-94-007-0896-9_13,

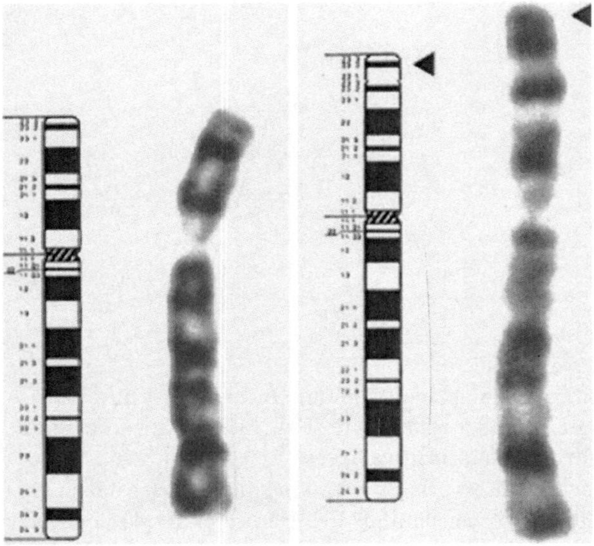

Fig. 13.1 Normal chromosome 8 and ideogram (*left*), and chromosome 8 and ideogram (*right*) with duplication of 8p23.2 (*arrowhead*) and parts of bands 8p23.1 and 8p23.3. FISH studies confirmed the extra band to be from chromosome 8. [Contributor: J. J. M. Engelen, Universities Maastricht, The Netherlands (c40)]

References

1. Lubs HA, Patil SR, Kimberling WJ, Brown J, Cohen M, Gerald P, Hecht F, Myrianthopoulos N, Summit RL (1977) Q and C-banding polymorphisms in 7 and 8 year old children: racial differences and clinical significance In: Hook EB, Porter IH (eds) Population cytogenetic studies in humans. Academic, New York, pp 133–159
2. McKenzie WH, Lubs HA (1975) Human Q and C chromosomal variations: distribution and incidence. Cytogenet Cell Genet 14:97–115
3. Jalal SM, Ketterling RP (2004) Euchromatic Variants. In: Wyandt HE, Tonk VS (eds) Atlas of human chromosome heteromorphisms. Kluwer Academic, Dordrecht, pp 75–95
4. Barber JCK, Joyce CA, Collinson MN, Nicholson JC, Wilatt LR, Dyson HM, Bateman MS, Green AJ, Yates JRW, Dennis RD (1998) Duplication of 8p32.1: a cytogenetic anomaly with no established clinical significance. J Med Genet 35:491–496
5. Engelen JJM, Moog U, Evers JLH, Dassen H, Albrechts JCM, Hamers AJH (2000). Duplication of chromosome region 8p23.1→p23.3. Am J Med Genet 91:18–21
6. Ludecke H, Johnson C, Wagner M et al (1991) Molecular definition of the shortest region of deletion overlap in the Langer-Giedion syndrome. Am J Hum Genet 49:1197–1206
7. Batanian JR, Morris K, Ma E, Huang Y, McComb J (2001) Familial deletion of (8)(q24.1q24.22) associated with a normal phenotype. Clin Genet 60:371–373
8. Fennel SJ, Benson JW, Kindley AD, Schwarz MJ, Czepulkowski B (1989) Partial deletion 8q without Langer-Giedion syndrome: a recognizable syndrome. J Med Genet 26(3):167–171

Chapter 14
Chromosome 9

The lightly-stained secondary constriction of chromosome 9 by plain Giemsa distinguishes it from other C-group chromosomes [1–3]. By G-banding, the secondary constriction itself stains lightly, but bands on either side of the centromere typically stain as intensely as the pericentromeric regions of other chromosomes (Fig. 14.1a–i). By C-banding or DA/DAPI staining, the entire region, including the pericentromeric bands usually consists of a uniform block of dark or brightly staining heterochromatin (Fig. 14.1m, n). Sometimes, however, blocks of heterochromatin will appear to be separated by a G-positive, C-negative band (Fig. 14.1j, k). More rarely, such C-negative bands can be quite striking and have been the object of considerable study (see Euchromatic Variants). The 9qh region is also strikingly stained by Giemsa-11 (Fig. 2.4) [4, 5].

While striking variations in the length of the 9qh region were frequently noted, it was not always clear whether or not such variations were innocuous. Lubs et al. [6] reported a frequency of 20–25% with 9qh variants in 7 and 8 year olds. While the frequency of 9qh+ was highest in black children with mental retardation (17%) the frequency in intellectually normal white children was higher (9%) than in white children with mental retardation (2%). Funderburk et al. [7], studying minor variants in psychiatric subjects, found an increased frequency of 9qh+ but only a random association with other prominent h regions, i. e. prominent satellites or a long Y. Soudek et al. [8] studied white male patients with idiopathic mental retardation and normal male controls by Q- and C-banding. Q-banding only was done on an additional cohort of mentally retarded males and females who were not part of the controlled study. The results showed higher frequencies of 9qh– and 9qh+ in the retarded individuals, statistically significant for 9qh–, but of borderline significance for 9qh+. A higher frequency of 9qh+ was found in the additional cohort studied by Q-banding, but no uniform phenotype was associated. Other studies of the segregation of variant 9s in families with abnormal probands [9, 10] concluded that the 9qh+ regions were normal variants.

Variants of 9qh have also been examined with regard to infertility. Ford and Lester [11] studied frequencies of aneuploid cells and striking 9qh variants in small cohorts of fertile and subfertile men and in groups of females who were fertile but

H. E. Wyandt, V. S. Tonk, *Human Chromosome Variation: Heteromorphism and Polymorphism,* DOI 10.1007/978-94-007-0896-9_14,

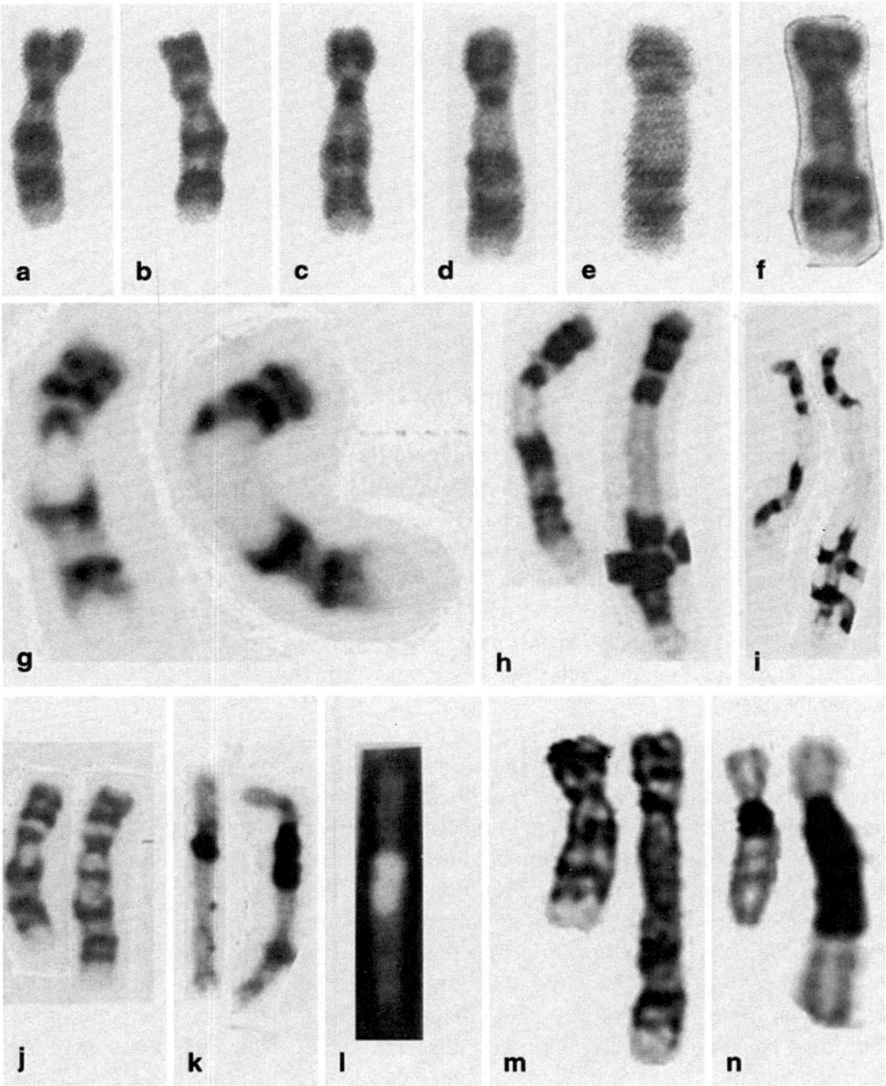

Fig. 14.1 **a–f** Chromosomes 9 showing variations in length by G-banding of the heterochromatic secondary constriction (9qh region). **g–i** Pairs of chromosomes 9 from the same subject at three different resolutions of G- banding. Note difference in size between the homologs in all three pairs. **j–l** Pairs of chromosomes 9 from the same subject showing a G-positive band in the right-hand homolog (**j**) and two blocks of heterochromatin by C-banding (**k**) and by DA/DAPI staining (**l**). **m, n** Pairs of 9s from normal male showing large 9qh region by G-banding (m) and C-banding (**n**). [Contributors: **a–e** Center for Human Genetics, Boston University (c1); **f** Jacqueline Schouman, Haukeland University Hospital (c25); **g–l** Lauren Jenkins, Kaiser Permanente (c2); **m, n** Jim Malone, Akron Childrens Hospital (c39)]

had histories of reproductive wastage. Men with inv(9) and 9qh+ had significantly increased frequencies of hyperdiploid cells. Females with 9qh+ and inv(9qh) also showed significant differences in rates of aneuploidy involving C through G groups with no preference for any particular group. Eiben et al. [12] found 9qh+ in 25% of teratozoospermic males who had a significantly higher fraction of malformed spermatozoa than a chromosomally normal group. A study of couples by Del Porto et al. [13] found inv(9) to be marginally related to recurrent miscarriages in an Italian population.

9qh inversions: A frequent variant of the 9qh region, besides length, is the apparent presence of a partial or complete inversion (Fig. 14.2). The frequency of partial 9qh inversions is 0.45–0.55% and of complete 9qh inversions, 0.13–1.07% [14–16]. Complete inversions are more frequent in the African-American than in the Caucasian population. Hansmann [17] studied the variability of 9qh inversions and classified them into three structural types: Type I showing total heterochromatin in the short arm, Type II showing part of the heterochromatin in the short arm, and Type III showing most of the heterochromatin in the long arm. Homozygous carriers of inversion 9 have been reported in at least four cases without apparent phenotypic effects [18].

Metaxotou et al. [19] studied polymorphism of chromosome 9 in 600 Greek subjects that included individuals with sex problems, congenital abnormalities and mental retardation, couples with multiple abortions and stillbirths, parents of children with Down syndrome or other malformations and relatives of affected individuals. Four percent (4%) of individuals carried a partial inversion of 9qh, 4% had complete inversions and 7.5% had a 9qh+ region. Howard-Peebles and Stoddard [20] reviewed the literature on 9qh inversions and outlined five possible effects: (1) no effect; (2) reduced fertility or reproductive failure; (3) duplication/deletion in meiosis; (4) chromosome interactions; (5) disturbances of RNA synthesis during meiosis. Most data supported (1). Of cases with reduced fertility, most were reported to be males. However, meiotic studies in two male carriers showed no abnormalities. Teo et al. [21] retrospectively, studied the significance of inv(9qh) in 2,448 antenatal patients and 1058 peripheral blood karyotypes. Twenty-nine families with inv(9qh) were found in the antenatal group and 6 cases were found in the peripheral blood group. Parents of the antenatal group contributed equally and inv(9qh) was not preferentially found in either fetal sex. All of the babies with inv(9qh) were phenotypically normal. Of the six cases from peripheral blood, one had tri(21) and one had del(13q). Subfertility appeared to be high (36%) among patients with inv(9qh), but this was interpreted to be due to a bias of older subfertile women seeking antenatal diagnosis. Interchromosomal or meiotic effects due to inv(9qh) or 9qh+ have been suggested [22–25]. However, because their frequencies in the population are so high, such occurrences are likely coincidental.

One recent family with inv(9)(p11q13) in grandmother and daughter is of interest in that the daughter who had muscle weakness and fatigue has a son who lacks inv(9) but has severe muscle disease with arthrogyposis (multiple contractures).

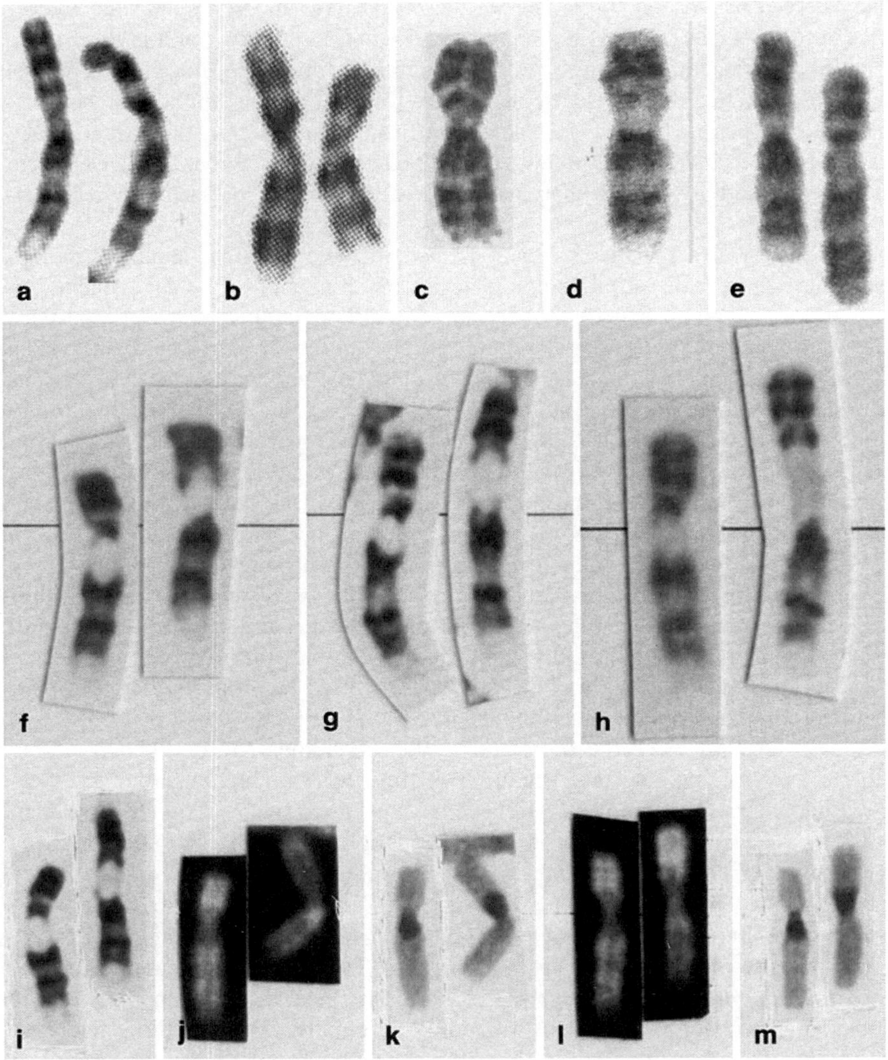

Fig. 14.2 **a–m** Pairs of chromosomes 9 (except for **c** and **d**) showing various pericentric inversions in one of the homologs. An example of a partial inversion is shown (**b**, *left*). All remaining pairs appear to be complete inversions. **i–k** G-,Q- and C- banding of pairs of homologs from the same subject. **l, m** Q and C banding of chromosomes 9 from a different subject. Inversions are on the left in each pair of homologs. [Contributors: **a, d, e** Center for Human Genetics, Boston University (c1); **b** Sharon Wenger, West Virginia University (c41); **c** Jacqueline Schoumans, Haukeland University Hospital (c22); **f–m** Lauren Jenkins, Kaiser Permanente (c2)]

A myopathy-associated tropomyosin 2 gene is at 9q13 (http://www.ncbi.nlm.nih.gov/omim/19099), raising speculation that a rearrangement in mother's 9qh region, could have resulted in a recombinant motif being transmitted to her son (Vijay Tonk and Golder Wilson, personal communication). Further evidence is needed, however, to support this speculation.

Structure of repeated sequences in 9qh: Satellite DNAs were initially isolated by ultracentrifugation in cesium chloride or cesium sulfate gradients and localized to the centromeric and secondary constriction regions in human chromosomes [26–32]. Satellite I hybridized mainly to distal Yq with minor concentrations in centromeres of other chromosomes, including chromosome 9. Satellite II hybridized mainly to chromosomes 1 and in low amounts to chromosome 16, but not to chromosome 9; satellite III hybridized mainly to chromosomes 9 with smaller amounts to the D- and G-group chromosomes; satellite IV hybridized to chromosomes 9 and the Y with smaller amounts on the D- and G-group chromosomes. These fractions of satellite DNA have since come to be referred to as "classical satellite" DNA. Additional satellite fractions were isolated [33, 34] using Hoechst 33258 and *EcoR1* restriction enzyme fractionation. The *EcoR1* fraction, referred to as alpha (α) satellite DNA, hybridizes to the centromeric regions of every chromosome [35]. An additional fraction of satellite DNA, referred to as beta (β) satellite hybridizes mainly to the centromere and short arm regions of the acrocentric chromosomes [36], but a specific fraction also hybridizes to the centromeric region of chromosome 9.

The molecular characterization of the 9qh region represents a challenging example of attempts to understand the organization and significance of heterochromatic regions. Because of the variability in size and frequency of structural rearrangements involving the 9qh region and the location of various satellite DNA fractions within it, there has been much interest in the molecular characterization, especially of inv(9qh), by *in situ* hybridization. Luke et al. [37] studied pericentric inversions in 9qh in five individuals using (1) a 171 bp α-satellite sequence that hybridized to all human centromeres, (2) a classical (satellite III) probe [D9Z1, pentameric repeat (5'ATTCC 3')] specific to chromosome 9qh, and (3) a β-satellite probe (D9Z5, 68 bp tandem repeat) specific to the centromeric region of chromosome 9. Four individuals had inverted alphoid sequences while one had a single intact alphoid site. Ramesh and Verma [38] studied breakpoints in α-, β- and satellite III sequences of inv(9qh) in an additional eight individuals. Three similar appearing inversions by CBG banding were classified by hybridization signal pattern (Figs. 14.3, 14.4) as: Type A (the most frequent) with two α-, one β- and one satellite III; Type B with two β-, one α- and one sat III; Type C, more complex, appeared to involve two inversions. Samonte et al. [39] describe eight more 9qh inversions with four additional types of α-, β- and satellite III signal patterns involving: (1) breaks within the α- and β-satellite regions; (2) breaks within the β-region; (3) breaks within the β- and satellite III regions; and (4) breaks within the α- and satellite III regions. Some of the latter cases are not as evident as having pericentric inversions at the microscope level.

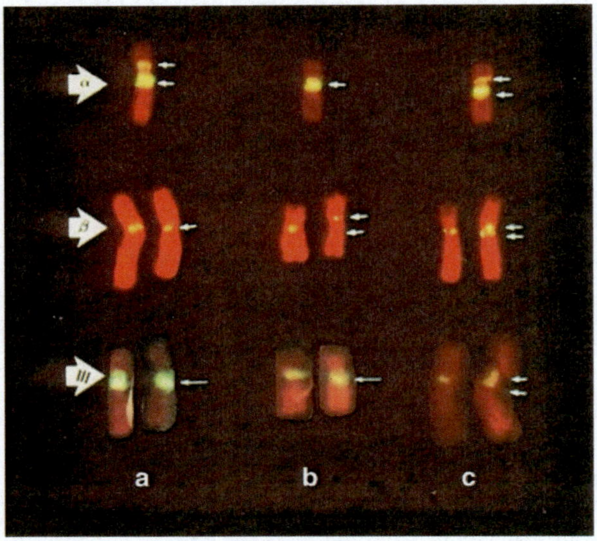

Fig. 14.3 Three different types of pericentric inversions involving shuffling of satellite DNAs within the 9qh region. [Reproduced with permission from Ramesh KH and Verma R (1996). J Med Genet 33: 395–398]. Characterization of pericentric inversion by FISH. **a** Arrows indicate the two alphoid (*top*), one β (*middle*), and one satellite III (*bottom*) hybridization signal(s); inversion type A (five cases). **b** Arrows indicate one alphoid (*top*), two β (*middle*), and one satellite III (*bottom*) hybridization signal(s); inversion type B (two cases). **c** Arrows indicate two alphoid (*top*), two β (*middle*), and two satellite III (*bottom*) hybridization signal(s); inversion type C (one case). The normal non-rearranged chromosome 9 is shown on the left for the chromosomes hybridized with β and satellite III DNA probes

The outcome of these studies is that there is considerable reshuffling of repeated sequences in the 9qh region. One question raised from these studies was "Is there formation of any functional dicentric in these reshufflings?" More recent work suggests that alphoid sequences themselves are not the requirement for a functional centromere. Vance et al. [40], for example, have reported a stable, apparently acentric chromosome derived from chromosome 9 without detectable α- or β-satellite DNA.

Euchromatic Variants: Structural variability of C-band-positive constitutive heterochromatin in chromosome 9 due to length variations or to partial and complete pericentric inversions has been discussed. However, G-positive, C- negative bands (Figs. 14.5a–e, 14.7, 14.8) can occur both within and outside these regions as apparent benign variants. Such bands have been described in both the long and short arms (Table 5.1, Chapter 5).

An inherited, extra G-positive band in the short arm of chromosome 9, proximal to the centromere, was first reported by Buckton et al. [41]. Two families with an extra G-positive band in the short arm of chromosome 9 in normal family members,

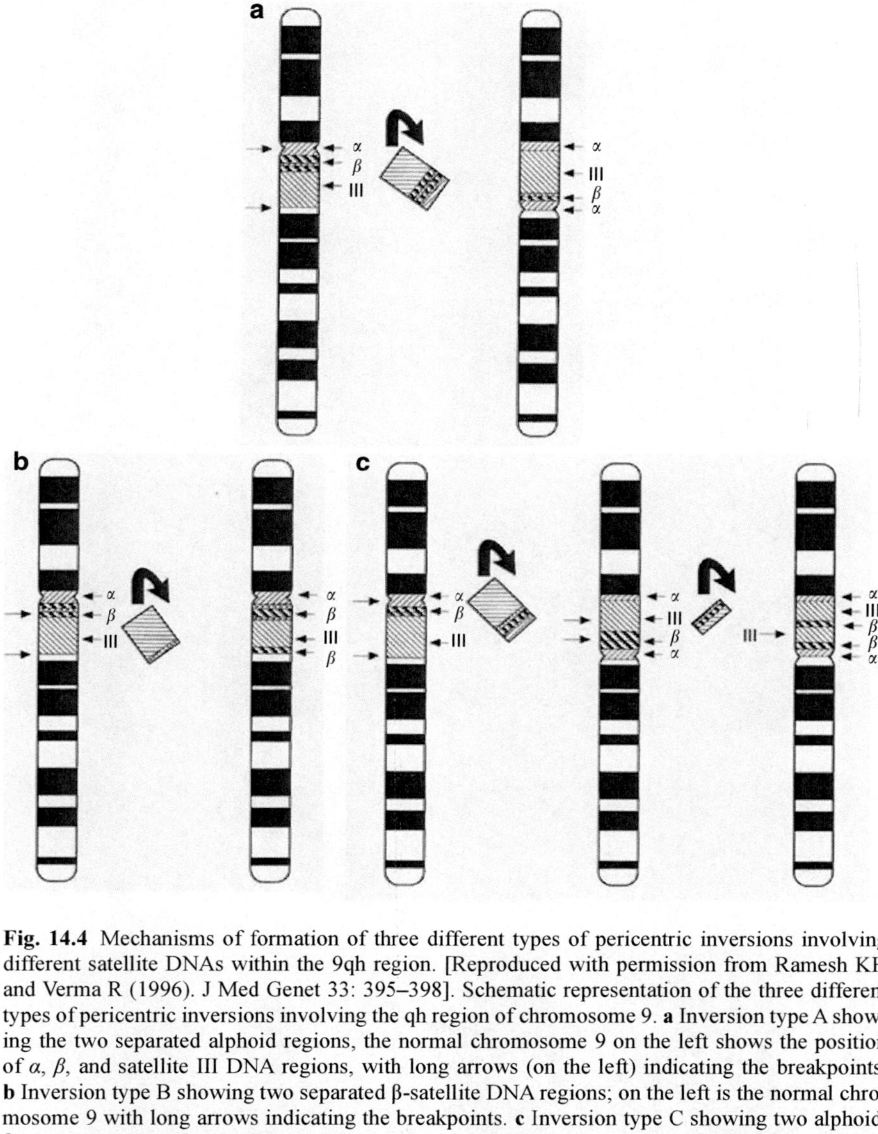

Fig. 14.4 Mechanisms of formation of three different types of pericentric inversions involving different satellite DNAs within the 9qh region. [Reproduced with permission from Ramesh KH and Verma R (1996). J Med Genet 33: 395–398]. Schematic representation of the three different types of pericentric inversions involving the qh region of chromosome 9. **a** Inversion type A showing the two separated alphoid regions, the normal chromosome 9 on the left shows the position of α, β, and satellite III DNA regions, with long arrows (on the left) indicating the breakpoints. **b** Inversion type B showing two separated β-satellite DNA regions; on the left is the normal chromosome 9 with long arrows indicating the breakpoints. **c** Inversion type C showing two alphoid, β, and satellite III DNA regions; breakpoints involved in the first and second inversions are shown by the long arrows

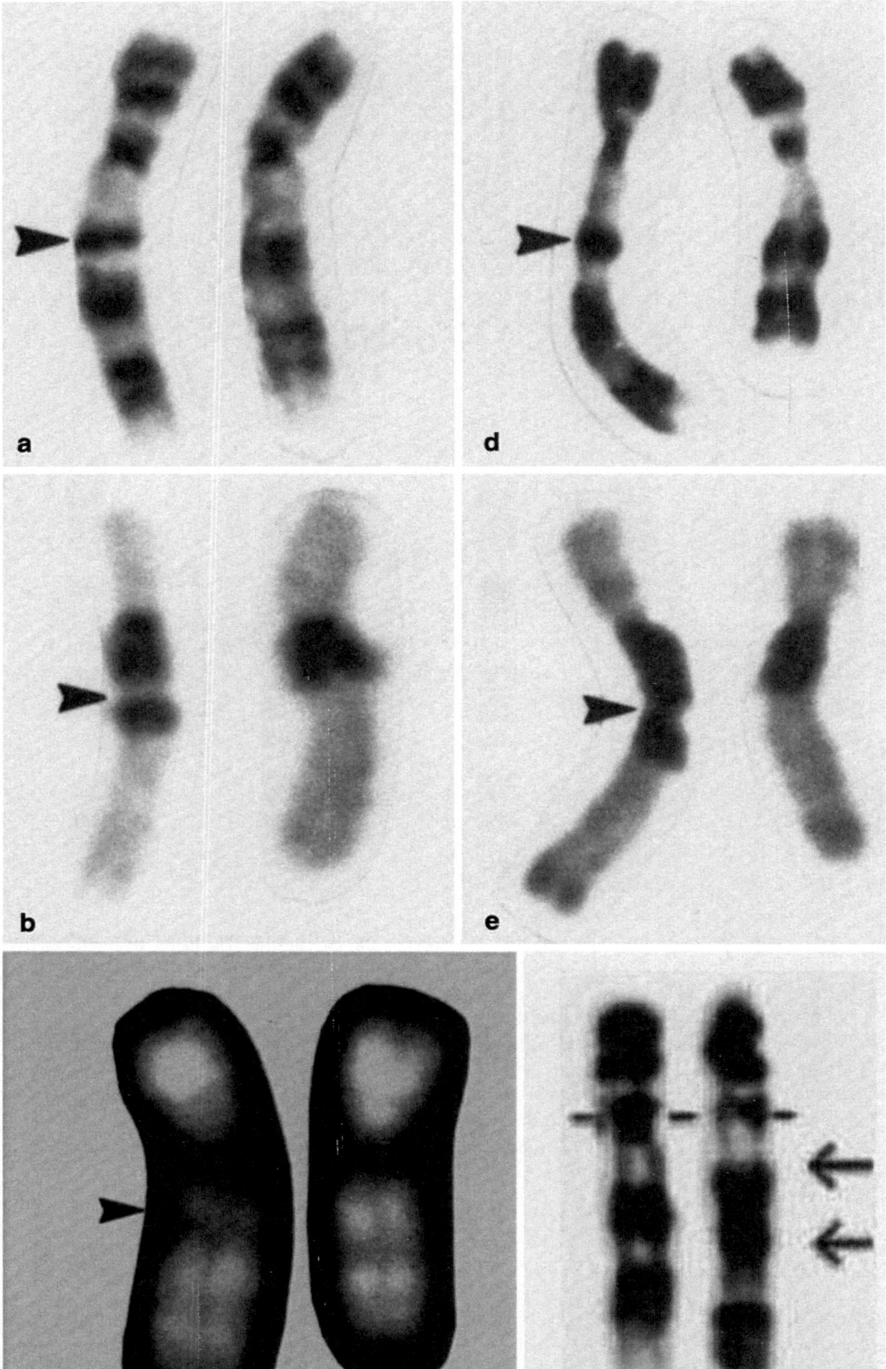
a
d
b
e
c
f

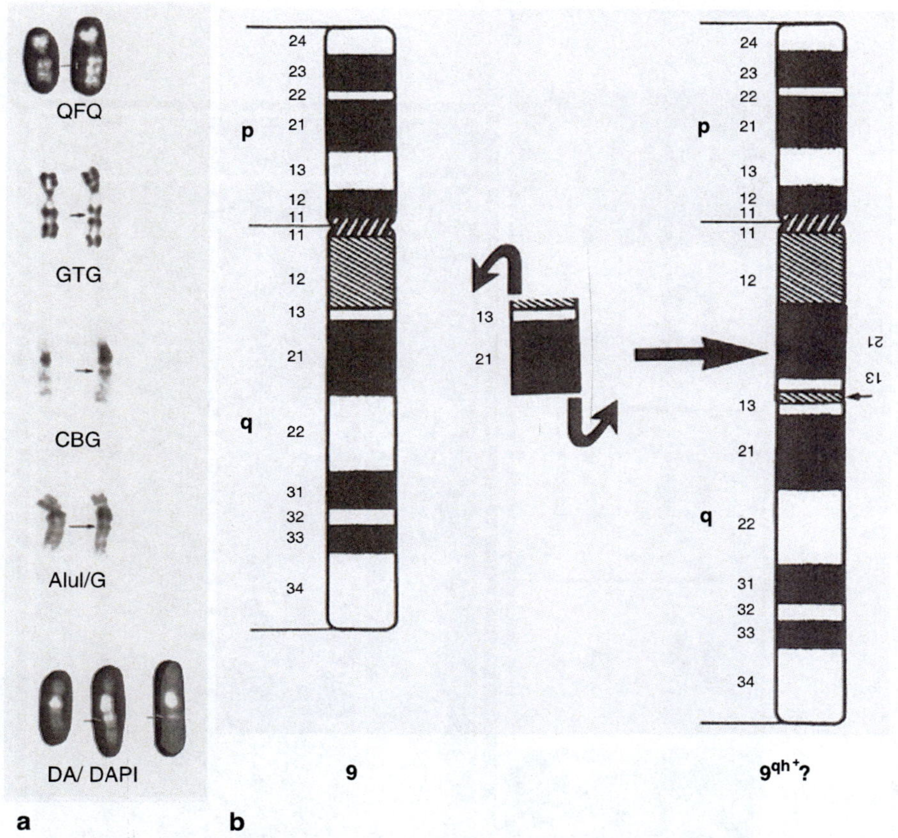

Fig. 14.6 **a** Extra euchromatic band in chromosome 9 in a 16.5 year-old female with mental retardation and multiple congenital abnormalities. Mother's chromosomes were normal; father was unavailable for study. **b** Diagramatic representation of proposed rearrangement giving rise to the extra band. [Reproduced with permission from Luke S et al. (1991). Am J Med Genet 40: 57–60]

in three generations were subsequently reported [42]. In one family, the extra band was C- positive and, therefore, not a euchromatic variant. However, in the other family the extra band was C-negative.

Jalal et al. [43] report a particularly large duplication of 9q13–q21 in a phenotypically normal mother and a fetal hydrops (Fig. 14.5f). A similar duplication was reported in a phenotypically normal baby boy and his mother [44]. Other families are reported with C-negative, G- or Q-positive bands within the 9qh

Fig. 14.5 **a–e** Pairs of chromosomes 9 showing extra euchromatic (G-positive) band (**a**, **d**), that is C-band negative (*Arrowhead*, b and e), and Q-band positive (*Arrowhead*, **c**). The extra band was found segregating as a normal variant in this family. A similar extra band was segregating in a second family. [Contributor: **a–e** Birgitte Roland, University of Calgary (c30)]. **f** Duplication of 9q13→9q21 in a phenotypically normal mother and a fetus with fetal hydrops. [Reproduced with permission from Jalal, SM, Kukolich MK, Garcia M, Day DW (1990). Euchromatic 9q+ heteromorphism in a family. Am J Med Genet 37: 155–156]

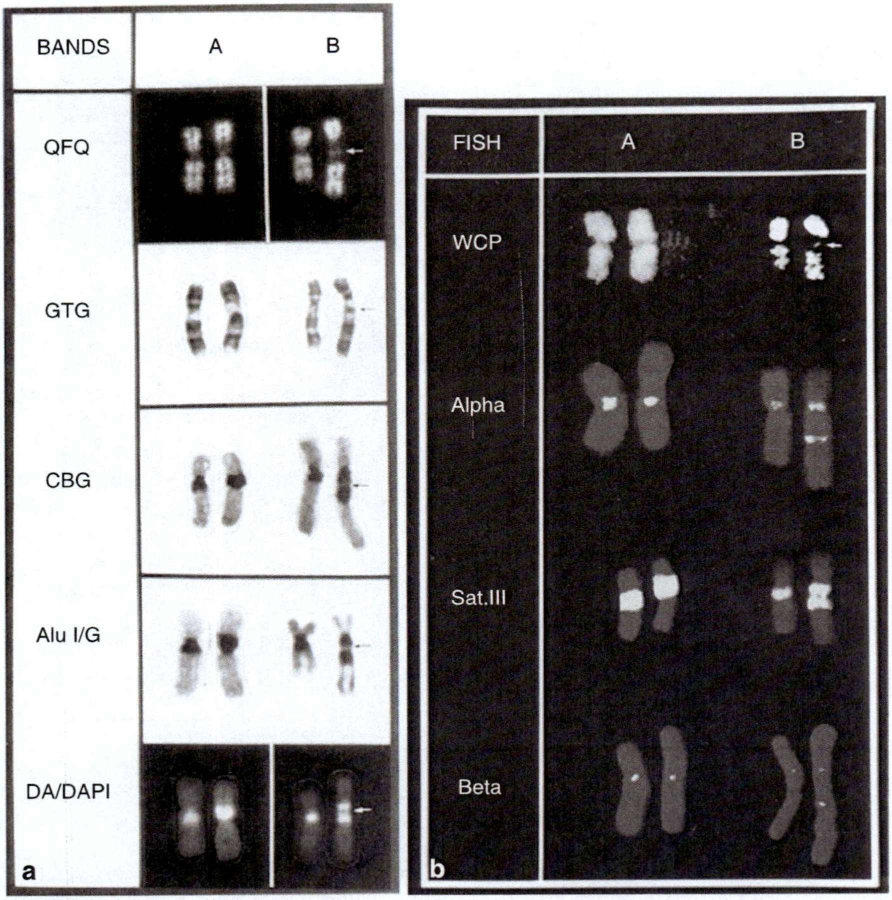

Fig. 14.7 Inversion in chromosome 9 (**a**), and extra euchromatic band (**b**), in two patients studied by multiple staining techniques (**a**), and by FISH (**b**). [Reproduced from Verma RS et al. (1993). Am J Hum Genet 52: 981–986]

region that were transmitted from normal fathers to a normal newborn daughter and to a normal fetus, respectively [45]. Verma et al. [46] speculated that an extra *de novo* G-positive euchromatic band within the 9qh region, in one of phenotypically normal twins originated from maternal band 9p12 during meiosis due to a large qh inversion, and that the extra band became "inactivated" when sandwiched within the heterochromatic 9qh region. Wang and Miller [47] and others [48] determined that some G-positive, C-negative bands were extra alpha satellite bands that possibly arose either from unequal crossing over or from an inversion that split the centromere. They also speculated that some euchromatic bands inserted into 9qh heterochromatin could be gene sequences that were inactivated by position effect.

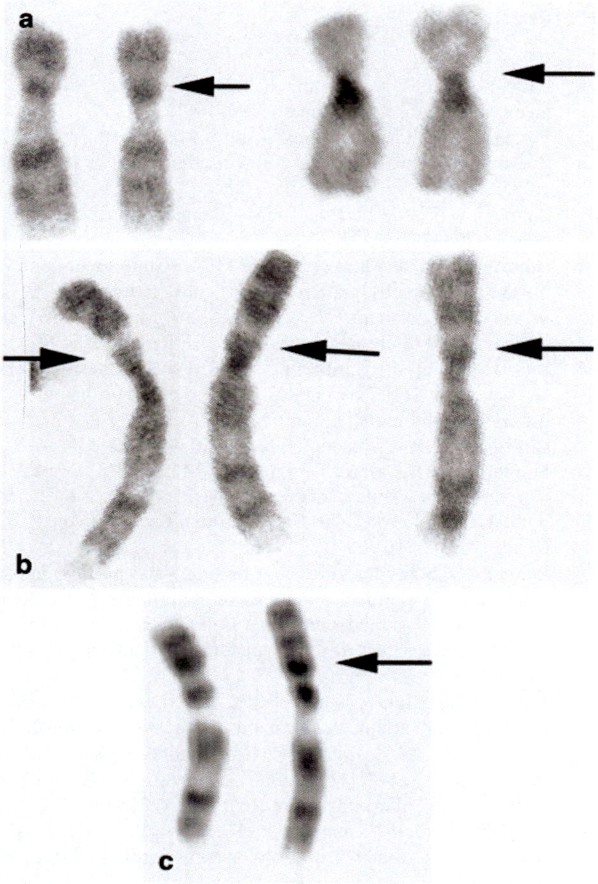

Fig. 14.8 Extra euchromatic bands in the short arm of chromosome 9. **a, b** Partial pericentric inversions of C-band-positive material in two normal amniotic prenatal cases. Right-hand pair of chromosomes in (**a**) is C-banded. [Contributor: Center for Human Genetics, Boston University (c1)]. **c** Pair of 9s representing a familial extra euchromatic band in the short arm of a mildly retarded 9 year-old girl and her father who also had learning disability and poor school performance. Mild dysmophic features were also present in both. [Reproduced from Haddad BR et al. (1996). J Med Genet 33: 1045–1047]

Some cases with extra G- or Q-positive bands inserted into the 9qh region have associated phenotypic abnormalities. Silengo et al. [49] reported a normal father and a 7-month old child with multiple congenital anomalies, both with an extra Q-positive band in the 9qh region. Luke et al. [50] reported a paracentric inversion and duplication in a 16.5 year-old female who has mental retardation and behavioral problems, which they regarded to be a structural abnormality (Fig. 14.6). However, Docherty and Hulten [51–52], reported a similar appearing chromosome 9 with a large euchromatic region inserted into the qh region in a child with trisomy 21. The same chromosome 9 was present in the phenotypically normal mother and a normal brother. An extra G-positive, C-negative band in 9p is frequently observed in normal persons (Fig. 14.8a,b). However, an extra G-positive, satellite III-negative band in the short arm of a chromosome 9 was reported in a father and his 9-year-old daughter, both with mild learning disabilities and dysmorphic features, including clinodachtyly (Fig. 14.8c) [53].

References

1. Patau K, Smith DW, Therman E, Inhorn SL (1960) Multiple congenital anomalies caused by an extra chromosome. Lancet 1:790–793
2. Ferguson-Smith MA, Ferguson-Smith ME, Ellis OM, Dickson M (1962) The site and relative frequencies of secondary constrictions in human somatic chromosomes. Cytogenetics 1:325–343
3. Palmer CG, Funderburk S (1965) Secondary constrictions in human chromosomes. Cytogenetics 4:261–276
4. Bobrow M, Madan K, Pearson PL (1972) Staining of some specific regions of human chromosomes, particularly the secondary constriction of No. 9. Nature New Biol 238:122–124
5. Wyandt HE, Wysham DG, Minden SK, Anderson RS, Hecht F (1976) Mechanisms of Giemsa banding of chromosomes. Exp Cell Res 102:85–94
6. Lubs HA, Patil SR, Kimberling WJ, Brown J, Cohen M, Gerald P, Hecht F, Myrianthopoulos N, Summit RL (1977) Q and C-banding polymorphisms in 7 and 8 year old children: racial differences and clinical significance. In: Hook EB, Porter IH (eds) Population cytogenetic studies in humans. Academic, New York, pp 133–159
7. Funderburk SJ, Guthrie D, Lind RC, Müller HM, Sparkes RS, Westlake JR (1978) Minor chromosome variants in child psychiatric patients. Am J Med Genet 1(3):301–308
8. Soudek D, Sroka H (1979) Chromosomal variants in mentally retarded and normal men. Clin Genet 16:109–116
9. Palmer CG, Schroder J (1971) A familial variant of chromosome 9. J Med Genet 8:202–208
10. Fitzgerald PH (1973) The nature and inheritance of an elongated secondary constriction on chromosome 9 of man. Cytogenet Cell Genet 12:404–413
11. Ford JH, Lester P (1978) Chromosomal variants and nondisjunction. Cytogenet Cell Genet 21:300–303
12. Eiben B, Leipoldt M, Rammelsberg O, Krause W, Engel W (1987) High incidence of minor chromosomal variants in teratozoospermic males. Andrologia 19(6):684–687
13. Del Porto G, D'Alessandro E, Grammatico P, Coghi IM, DeSanctis S, Giambenedetti M, Vaccarella C, Fabi R, Marciano MF, Nicotra M (1993) Chromosome heteromorphisms and early recurrent abortions. Hum Reprod 8(5):755–758
14. Hsu LYF, Benn PA, Tannenbaum HL, Perlis TE, Carlson AD (1987) Chromosome polymorphism of 1, 9 16 and Y in 4 major ethnic groups: a large prenatal study. Am J Med Genet 26:95–101
15. Boue J, Taillemite JC, Hazael-Massieux P, Leonard C, Boue A (1975) Association of pericentric inversion of chromosome 9 and reproductive failure in ten unrelated families. Humangenetik 30:217–224
16. Madan K, Bobrow M (1974) Structural variation in chromosome no. 9. Annals Genet 17:81–86
17. Hansmann I (1976) Structural variability of human chromosome 9 in relation to its evolution. Hum Genet 31:247–262
18. Vine DT, Yarkoni S, Cohen MM (1976) Inversion homozygosity of chromosome no. 9 in a highly inbred kindred. Am J Hum Genet 28:203–207
19. Metaxotou C, Kalpini-Mavrou A, Panagou M, Tsengi C (1978) Polymorphism of chromosome 9 in 600 Greek subjects. Am J Hum Genet 30:85–89
20. Howard-Peebles PN, Stoddard GR (1976) A satellited Yq chromosome associated with trisomy 21 and an inversion of chromosome 9. Hum Genet 34:223–225
21. Teo SH, Tan M, Knight L, Yeo SH, Ng I (1995) Pericentric inversion 9 –incidence and clinical significance. Ann Acad Med, Singapore. 24:302–304
22. Schinzel A, Hayashi K, Schmid W (1974) Mosaic trisomy and pericentric inversion of chromosome 9 in a malformed boy. Humangenetik 25:171–177
23. Bowen P, Ying KL, Chung GSH (1974) Trisomy 9 mosaicism in a newborn infant with multiple malformations. J Pediat 85:95–97

24. Seabright M, Gregson M, Mould S (1976) Trisomy 9 associated with an enlarged 9qh segment in a liveborn. Hum Genet 34:323–325
25. Sutherland GR, Gardiner AJ, Carter RF (1976) Familial pericentric inversion of chromosome 19, inv(19)(p13q13) with a note on genetic counseling of pericentric inversion carriers. Clin Genet 10:54–59
26. Saunders GF, Hsu TC, Getz MJ, Simes EL, Arrighi F (1972) Locations of human satellite DNA in human chromosomes. Nature New Biology 236:244–246
27. Jones KW, Corneo G (1971) Location of satellite and homogeneous DNA sequences on human chromosomes. Nat New Biol 233(43):268–271
28. Ginelli E, Corneo G (1976) The organization of repeated DNA sequences in the human genome. Chromosoma (Berl) 56:55–69
29. Buhler EM, Tsuchimoto T, Jurik LP, Stalder GR (1975) Satellite DNA III and alkaline Giemsa staining. Humangenetik 26:329–333
30. Corneo G, Ginelli E, Polli EJ (1968) Isolation of complementary strands of a human satellite DNA. J Mol Biol 33:331
31. Miklos GLG, John B (1979) Heterochromatin and satellite DNA in man: Properties and prospects. Am J Hum Genet 31:264–280
32. Gosden JR, Mitchell AR, Buckland RA, Clayton RP, Evans HJ (1975) The location of four human satellite DNAs on human chromosomes. Exp Cell Res 92:148–158
33. Manuelidis L, Wu JC (1978) Homology between human and simian repeated DNA. Nature 276:92–94
34. Manuelidis L (1978) Complex and simple sequences in human repeated DNAs. Chromosoma 66:1–22
35. Willard HF (1985) Chromosome-specific organization of human alpha satellite DNA. Am J Hum Genet 37:524–532
36. Waye JS, Willard HF (1989) Human α-satellite DNA: Genomic organization and sequence definition of a class of highly repetitive tandem DNA. Proc Natl Acad Sci U S A 86:6250–6254
37. Luke S, Verma RS, Conte RA, Mathews T (1992) Molecular characterization of the secondary constriction region (qh) of human chromosome 9 with pericentric inversion. J Cell Sci 103:919–923
38. Ramesh KH, Verma RS (1996) Breakpoints in alpha, beta and satellite III DNA sequences of chromosome 9 result in a variety of pericentric inversions. J Med Genet 33:395–398
39. Samonte RV, Conte RA, Ramesh KH, Verma RS (1996) Molecular cytogenetic characterization of breakpoints involving pericentric inversions of human chromosome 9. Hum Genet 98:576–580
40. Vance GH, Curtis CA, Heerema NA, Schwartz S, Palmer CG (1997) An apparently acrocentric marker chromosome originating from 9p with a functional centromere without detectable alpha and beta satellite sequences. Am J Hum Genet 71:436–442
41. Buckton KE, O'Riordan ML, Ratcliffe S, Slight J, Mitchell M, McBeath S, Keay AJ, Barr D, Short M (1980) A G-band study of chromosomes in liveborn infants. Ann Hum Genet 43:227–239
42. Sutherland GR, Eyre H (1981) Two unusual G-band variants of the short arm of chromosome 9. Clin Genet 19:331–334
43. Jalal SM, Kukolich MK, Garcia M, Day DW (1990) Euchromatic 9q+ heteromorphism in a family. Am J Hum Genet 37:155–156
44. Knight LA, Soon GM, Tan M (1993) Extra G positive band on the long arm of chromosome 9. J Med Genet 30:613
45. Roland B, Chernos JE, Cox D (1992) M. 9qh+ variant band in two families. Am J Med Genet 42:137–138
46. Verma RS, Luke S, Brennan JP, Mathews T, Conte RA, Macera MJ (1993) Molecular topography of the secondary constriction region (qh) of the human chromosome 9 with an unusual euchromatic band. Am J Hum Genet 52:981–986
47. Wang JC, Miller WA (1994) Molecular cytogenetic characterization of two types of chromosome 9 variants. Cytogenet Cell Genet 67:190–192

48. Macera MJ, Verma RS, Conte RA, Bialer MG, Klein VR (1995) Mechanisms of the origin of a G-positive band within the secondary constriction region of human chromosome 9. Cytogenet Cell Genet 69:235–239
49. Silengo MC, Davi GF, Franeschini P (1982) Extra band in the 9qh+ chromosome in a normal father and in his child with multiple congenital anomalies. Hum Genet 60:294
50. Luke S, Verma RS, PeBenito R, Macera M (1991) J. Inversion-duplication of bands q13–q21 of human chromosome 9. Am J Med Genet 40:57–60
51. Docherty Z, Hulten MA (1985) Extra euchromatic band in the qh region of chromosome 9. J Med Genet 22:156–157
52. Docherty Z, Hulten MA (1993) Rare variant of chromosome 9 (letter). Am J Med Genet 45:105–106
53. Haddad BR, Lin AE, Wyandt H, Milunsky A (1996) Molecular cytogenetic characterization of the first familial case of partial 9p duplication (p22p24). J Med Genet 33:1045–1047

Chapter 15
Chromosome 10

A common variant reported for chromosome 10 is a pericentric inversion, i.e. inv(10)(p11.2q21.2) (Fig. 15.1a, b). Collinson MN et al. [1] studied 33 families with a pericentric inversion of chromosome 10 and reviewed the literature for another 32 families. Of their own families, twenty one were ascertained through prenatal diagnosis and twelve were ascertained for a variety of reasons as were reported cases. From the compiled data, they concluded that reasons for referral were unrelated to the chromosome abnormality. There was no recorded instance of a recombinant 10 arising from inv(10). The rate of abortion from 94 pregnancies in their families was 7.4% (less than 15% for the general population). Stillbirth and neonatal death, corrected for ascertainment bias, was 1.3% (compared to 0.57% for the general population). Fertility also appeared not to be reduced in either male or female carriers. Overall, they conclude that inv(10) can be regarded as a variant analogous to the pericentric inversion in chromosome 2 and that investigation of carrier status in families with the inversion is unwarranted since there are no known consequences.

Isolation of a specific alpha satellite for chromosome 10 [2], consist of 8 tandem repeats of a 171 bp monomer unit. A cloned 8mer representative probe detects a polymorphic restriction enzyme pattern. Jackson et al. [3] studied a cosmid containing the junction between alphoid and satellite III sequences mapping to chromosome 10. They showed the alphoid sequences consisted of tandemly arranged dimers that were distinct from the known chromosome 10-specific alphoid family and placed the boundary between alphoid and satellite III in the 10 cen-10q11.2 region. Sequence data showed separation of the repetitive sequences by a partial L1 repeat sequence less than 500 bp in length.

H. E. Wyandt, V. S. Tonk, *Human Chromosome Variation: Heteromorphism and Polymorphism,* DOI 10.1007/978-94-007-0896-9_15,

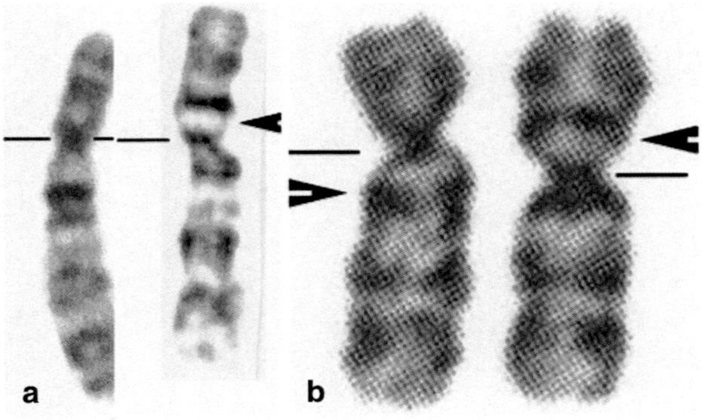

Fig. 15.1 **a** Pericentric inversion of chromosome 10 from prenatal sample. Inversion is maternal in origin. **b** Pair of chromosomes 10 from prenatal sample with pericentric inversion in chromosome at left. Parental studies were not done. [Contributors: **a** Center for Human Genetics, Boston University (c1); **b** K. Yelavarthi and J. Nunich, Northwest Center for Medical Research (c13)]

References

1. Collinson MN, Fisher AM, Walker J, Currie J, Williams L, Roberts P (1997) Inv(10)(p11.2q21.2), a variant chromosome. Hum Genet 101:175–180
2. Devilee P, Kievits T, Waye JS, Pearson PL, Willard HF (1988) Chromosome-specific alpha satellite DNA: isolation and mapping of a polymorphic alphoid repeat from chromosome 10. Genomics 3:1–7
3. Jackson MS, Mole SE, Ponder BA (1992) Characterization of the boundary between satellite III and alphoid sequences on human chromosome 10. Nucleic Acids Res 20:4781–4787

Chapter 16
Chromosome 11

A variant of the centromeric region of chromosome 11 has been observed in at least three cases. Till et al. [1] describe a duplication of the centromere in a fetus and a 39 year-old mother. The fetus had no obvious abnormalities. C- and G- banding suggested the variant chromosome was a pseudodicentric with inactivation of one centromere. A case with a large C-band positive, Q-band negative centromeric region of chromosome 11 (Fig. 16.1a) was observed in a man identified through fetal loss in his wife [2]. The authors reported this as a familial variant whose clinical significance is unknown. A third case of a large centromeric variant (Fig. 16.1b), ascertained prenatally, that was paternal in origin, and positive for alpha satellite D11Z1, is also know to us [3].

Euchromatic variant: Interstitial deletion of the G dark band at 11p12 has been reported in a mother, her normal son and a phenotypically normal female newborn [4].

H. E. Wyandt, V. S. Tonk, *Human Chromosome Variation: Heteromorphism and Polymorphism*, DOI 10.1007/978-94-007-0896-9_16,

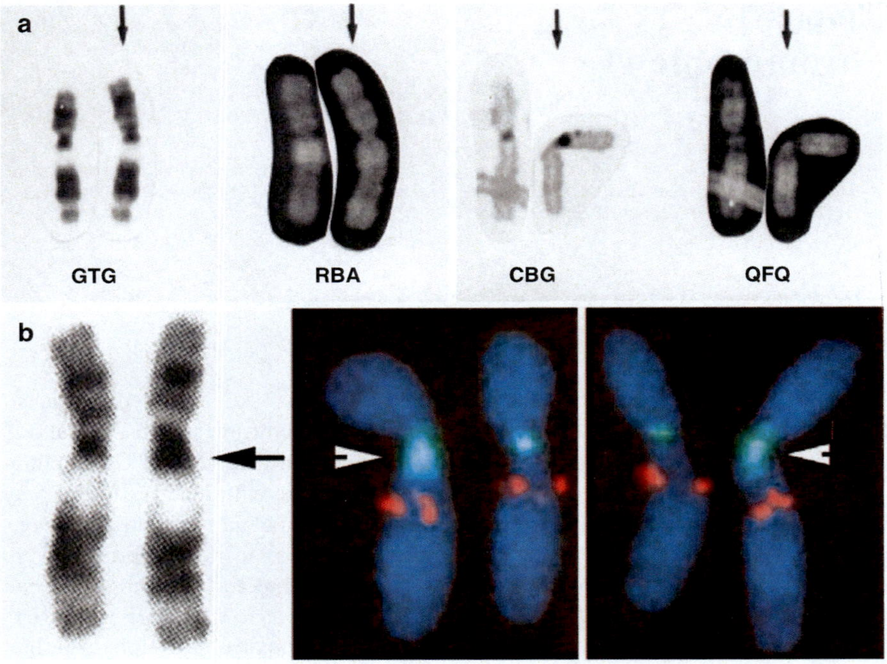

Fig. 16.1 **a** Chromosome 11 pairs showing prominent G-positive (GTG) and C-positive band in the short arm. Region is also R-positive (RBA) and Q-negative (QFQ). Arrows point to variant chromosome (*right*) in each pair. [Modified Fig. 1 reproduced with permission from Aiello V et al. (1994). Am J Med Genet 50:294–295]. **b** Pairs of chromosome 11: Left, by G-banding at left with arrow pointing to a large heterochromatic region. Middle and right, two pairs from same patient (*at right*) showing large alpha satellite signal (*arrows*) with D11Z1 probe (*green*). Variant chromosome was paternal in origin. Red signals are for cyclin D1 at 11q13. [Contributors: **b**. Patricia Miron, Brigham and Women's Hospital (c9)]

References

1. Till M, Rafat A Charrin C, Plauchu H, Germain D (1991) Duplication of chromosome 11 centromere in fetal and maternal karyotypes: a new variant? Prenat Diagn 11:481–482
2. Aiello V, Ricci N, Palazzi P, D'Agostino G, Calzolari E (1994) New variant of chromosome 11. Am J Med Genet 50:294–295
3. Wyandt HE (2004) Inroducton. In: Wyandt HE, Tonk VS (eds) Atlas of human chromosome heteromorphisms. Kluwer Academic, Dordrecht, pp 3–10
4. Barber JCK, Mahl H, Portch J, Crawfurd MDA (1991) Interstitial deletions without phenotypic effect: prenatal diagnosis of a new family and brief review. Prenat Diagn 11:411–416

Chapter 17
Chromosome 12

Mayer et al. [1] reported an enlarged centromere on a chromosome 12, ascertained in a patient with mental retardation, that was also present in many normal relatives. A similar heteromorphism was reported in three Japanese A-bomb survivors [2].

References

1. Mayer M, Matsuura J, Jacobs, P (1978) Inversions and other unusual heteromorphisms detected by C-banding. Hum Genet 45:43–50
2. Sofuni T, Tanabe K, Ohtaki K, Shimba H, Awa AA (1974) Two new types of C-band variants in human chromosome (6ph+ and 12ph+). Jpn J Hum Genet 19:251–256

H. E. Wyandt, V. S. Tonk, *Human Chromosome Variation: Heteromorphism and Polymorphism,* DOI 10.1007/978-94-007-0896-9_17,

Chapter 18
Chromosome 13

The most variable chromosomes in the human karyotype are the acrocentric chromosomes, 13, 14, 15, 21 and 22. The features in common are: (1) they all carry nucleolar organizing regions (NORs) revealed by silver staining; (2) they all have four distinct regions (satellite, stalk, short arm and centromere) (Fig. 18.1a) that can vary in size and/or may have different staining properties (Fig. 18.1b)[1]; (3) all are involved in Robertsonian translocations in which there is typically loss of satellites, NOR's and part or all of the short arms. The involvement of acrocentrics in Robertsonian translocations is non-random [2–4]. Translocation between chromosomes 13 and 14 is the most common, followed by translocation between 14 and 21, then between 13 and 21[4]. Such non-random involvement appears to be correlated with the frequency that the acrocentrics share repeated sequences [5, 6]. *In situ* hybridization studies show that the majority of Robertsonian translocations involving non-homologs are dicentric, whereas, the majority involving homologs are monocentric [7, 8]. Molecular studies reveal varying degrees of homology of repeated sequences, with breakpoints most frequently occurring in repeated sequences that are in common, typically in the short arm, less frequently in the NOR or satellite regions [9–12]. The common nucleolar organizing function that brings NOR regions and other shared repeated sequences in the acrocentric short arms in close proximity, undoubtedly contribute to the high frequency of Robertsonian rearrangement.

There is considerable shuffling of various repetitive DNA sequences comprising the pericentromeric, short-arm and satellites regions of these chromosomes. Variations include staining and/or size of the pericentromeric regions, short arms, and satellites (Fig. 18.2a–e) as well as variation in the number and/or size of NORs (Fig. 18.2f, g). None of these variations appear to have any direct clinical consequences. Occasionally, an acrocentric with a very small or absent short arm is seen. Although usually a normal variant, on rare occasion it may be symptomatic of a chromosome that is unstable [13] or has a cryptic translocation. Very short, intensely bright short arms are more frequent on chromosome 13 than chromosomes 14 or 15. Q-bright variants frequently correspond to intensely Giemsa-posi-

[1] Acridine orange staining was performed following treatment of chromosomes with dilute alkali (0.07M NaOH) to give RFA banding pattern [Wyandt et al. (1974). Humangenetik. 23:119–130] [1].

H. E. Wyandt, V. S. Tonk, *Human Chromosome Variation: Heteromorphism and Polymorphism,* DOI 10.1007/978-94-007-0896-9_18,

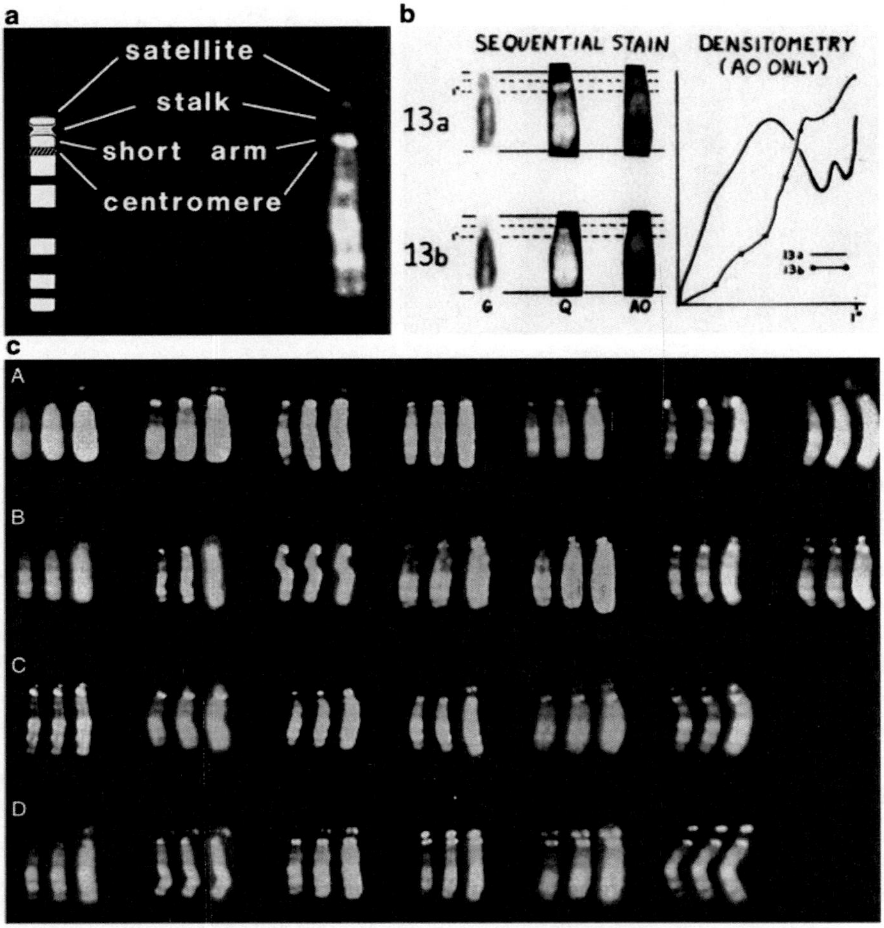

Fig. 18.1 **a** Ideogram and photograph of chromosome 13 illustrating independently varying regions of acrocentric chromosomes (Figure 1, reprinted with permission from Olson et al. (1986). Am J Hum Genet 38: 235–252). **b** Sequential banding of chromosome 13 homologs by G-, Q- and R- banding by acridine orange (AO) [1] followed by densitometry (*graph at right*) of the variable regions stained by AO. Note differences in intensity of Giemsa staining of Q- negative regions for the two homologs and additional variability in staining of both Q-positive and Q-negative regions by AO staining (H.E. Wyandt, unpublished). **c** Twenty-six variants of chromosome 13 from a population of 39 unrelated people. Each chromosome is serially printed to reveal heteromorphisms not visible at an exposure generally chosen to define the overall banding pattern. A1-A7 have satellites that would not have been observed at a routine exposure. Scores are determined by comparisons of serial prints against standards, including an internal standard. Very short, medium to intense short arms (scored 13, 14 and 15) as in C1 or D3 are relatively common on chromosomes 13, whereas bright short arms on chromosomes 14 and 15 are less frequent. [Reproduced from Olson SB et al., Am J Hum Genet 38:235–252 (1986)]

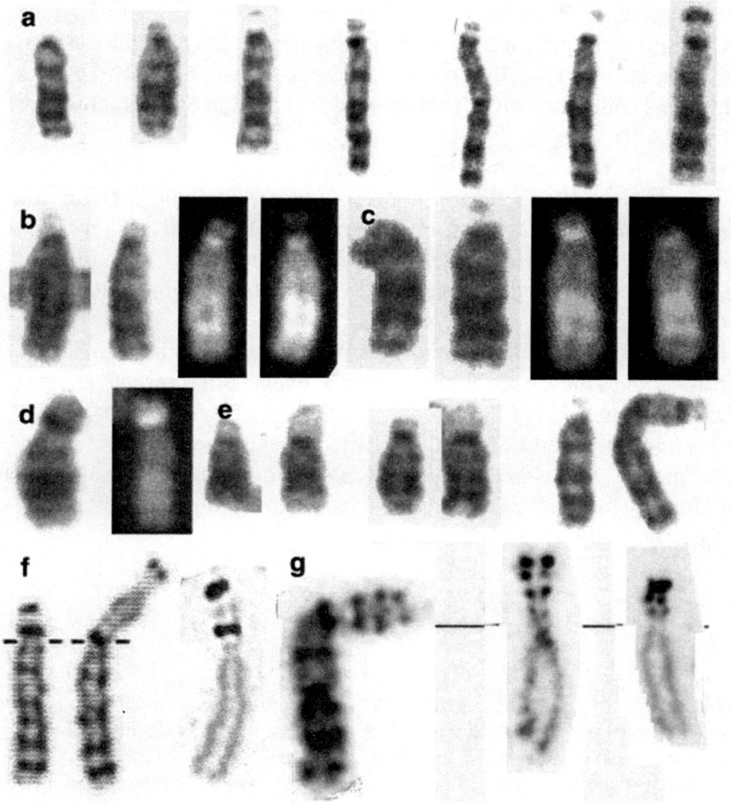

Fig. 18.2 **a** Normal chromosomes 13 from seven different individuals showing increasing band resolution and variations in size and staining of short arms, stalks and satellites by G-banding. Note the range from an almost undetectable short arm and apparent lack of a satellite (*extreme left*) to a more prominent dark staining short arm and very prominent satellite (extreme right). **b** Pairs of homologs (different cells from same subject) showing almost no evidence of satellites by G-banding, but clearly showing the presence of satellites by Q-banding. **c** Pairs of homologs from a different subject showing similarly staining short arms by G-banding, but with a striking difference in size and intensity by Q-banding. **d** Chromosome 13 with prominent light and dark staining regions in the short arm, with the prominent dark band being intensely bright by Q-banding (note a less intensely fluorescent satellite is also evident by Q-banding). **e** G-banding variations in the size of stalk regions (regions separating short arm and satellite) in pairs of homologs from three different individuals. **f** Pair of 13s from a normal individual showing striking stalk region by G-banding (middle) with an unusual pattern of alternating light (constrictions) and darker bands and a terminal satellite. NOR (silver)-staining (left) reveals two prominent nucleolar organizing regions (NORs) more or less corresponding to the two constricted regions. **g** G-banding (left) showing a chromosome 13 with 3 tandem dark staining satellites. NOR-staining (middle and right) of the same chromosome from two different cells showing at least three NOR regions. In such tandemly repeated NOR regions, the number of active NORs may vary in size and number in different cells. [Contributors: **a–e** Center for Human Genetics, Boston University (c1); **f** Arturo Anguiano, Quest Diagnostics (c17); **g** Lauren Jenkins, Kaiser Permanente (c2)]

tive regions by G-banding. However, the correlation is not definite. Dark G-bands are not always Q-bright. Furthermore, neither dark G band nor Q-bright regions, especially in the short arm and satellite regions, are necessarily C-band positive. Secondary constrictions containing NOR regions also do not stain by C-banding but are R-band positive [14, 15].

Because of the high degree of heteromorphism, variants of the acrocentric chromosomes have been particularly useful to (1) determine parental and meiotic origin of chromosomal aneuploidy; (2) determine the origins of triploidy, (3) rule out maternal contamination; 4) determine paternity (Fig. 18.1c) [16]. Q-banding for paternity testing [17] revealed twenty-six different variants of chromosome 13 in a population of 39 unrelated people.

In all cases of striking or unusual heteromorphisms, multiple banding techniques and parental chromosome studies are recommended, even if the variant appears similar to one that has been seen before.

Euchromatic Variant: Deletion of band 13q21 by chromosome analysis was reported in a son and mother who were both phenotypically normal [18]. However, a family history of multiple miscarriages raises the question of whether an undetected balanced rearrangement could have been present.

References

1. Wyant HE, Vlietinck RF, Magenis RE, Hecht F (1974) Colored reverse-banding of human chromosomes with acridine orange following alkaline-formalin treatment: densitometric validation and applications. Humangenetik 23:119–130
2. Hecht F, Case MP, Lovrien EW, Higgins JV, Thuiline HC, Melnyk J (1968) Non-randomness of translocations in man: preferential entry of chromosomes into 13-15-21 translocations. Science 161:371–372
3. Nagel M, Hoehn H (1971) On the non-random involvement of D-group chromosomes in centric fusion translocations in man. Humangenetik 11:351–354
4. Hecht F, Kimberling WJ (1972) Registry of data on Robertsonian (centric fusion) translocations. Lancet 1:1342
5. Jorgensen AL, Bostock CJ, Bak AL (1987) Homologous subfamilies of human alphoid repetitive DNA on different nucleolus organizing chromosomes. Proc Natl Acad Sci U S A 84:1075–1079
6. Choo KH, Vissel B, Brown R, Filby RG, Earle E (1988) Homologous alpha satellite sequences on human acrocentric chromosomes with selectivity for chromosome 13, 14 and 21: implications for recombination between nonhomologs and Robertsonian translocations. Nuc Acids Res 16:1273–1284
7. Wolff DJ, Schwartz S (1992) Characterization of Robertsonian translocations by using fluorescence *in situ* hybridization. Am J Hum Genet 50:174–181
8. Gravholt CH, Friedrich U, Caprani M, Jorgensen AL (1992) Breakpoints in Robertsonian translocations are localized to satellite III DNA by fluorescence *in situ* hybridization. Genomics 14:924–930
9. Page SL, Shin JC, Han JY, Choo KH, Shaffer LG (1996) Breakpoint diversity illustrates distinct mechanisms for Robertsonian translocation mechanisms. Hum Mol Genet 5:1279–1288
10. Sullivan BA, Jenkins LS, Karson FM, Leana-Cox J, Schwartz S (1996) Evidence for structural heterogeneity from molecular cytogentic analysis of dicentric Robertsonian translocations. Am J Hum Genet 59:167–175

11. Bandyopadhyay R, McQuillan C, Page SL, Choo KH, Shaffer LG (2001) Identification and characterization of satellite III subfamilies to the acrocentric chromosomes. Chromosome Res 9:223–233.
12. Bandyopadhyay R, Berend SA, Page SL, Choo KH, Shaffer LT (2001) Satellite III sequences on 14p and their relevance to Robertsonian translocation formation. Chromosome Res 9:235–242
13. Lebo RV, Wyandt HE, Warburton P, Li S, Milunsky JM (2002) An unstable dicentric Robertsonian translocation in a markedly discordant twin. Clin Genet 65:383–389
14. Balicek P, Zizka J (1980) Intercalar satellites of human acrocentric chromosomes as a cytological manifestation of polymorphisms in GC-rich material. Hum Genet 54:343–347
15. Balicek P, Zizka J, Skalska H (1982) RGH-band polymorphism of the short arms of human acrocentric chromosomes and relationship of variants to satellite association. Hum Genet 62:237–239
16. Olson SB, Magenis RE (2004) Technical variables and the use of heteromorphisms in the study of human chromosomes. In Wyandt HE, Tonk VS (eds) Atlas of human chromosome heteromorphisms. Kluwer Academic, Dordrecht, pp 63–73
17. Olson SB, Magenis RE, Lovrien EW (1986) Human chromosome variation: the discriminatory power of Q-band heteromorphism (variant) analysis in distinguishing between individuals, with specific application to cases of questionable paternity. Am J Hum Genet 38:235–252
18. Coutuier J, Morichon-Delvallez N, Dutrillaux B (1986) Deletion of band 13q21 is compatible with normal phenotype. Hum Genet 70:87–91

Chapter 19
Chromosome 14

As with chromosome 13, variations in the staining of centromeric and short arm regions are observed (Fig. 19.1). Q-band variants of chromosome 14 include bright satellites, bright short arms, lack of satellites or NORs, and variable size short arms. A bright short arm region seems to be less frequent than for chromosome 13. In a study of 39 unrelated people [1], fourteen different variants were distinguishable by Q-banding (Fig. 19.2).

Of particular interest with regard to NOR regions is a study of Balicek et al. [2, 3] in which they found that multi-satellited, multi-NOR-positive chromosomes 14 showed correspondingly large blocks of R-banded (RGH-positive) material that were heat resistant (Fig. 19.3 and 19.4). Treatment of lymphocytes with azacytidine in low doses led to extreme under condensation of RGH-positive regions. Intercalary NOR-positive blocks corresponding to secondary constrictions were shown to punctuate these regions anywhere along their length. From these observations they concluded that R-banded material represents genetically inactive blocks of rRNA genes.

FISH variants of chromosome 14. Fluorescent *in situ* hybridization has revealed at least one common variant on chromosome 14 that shows up when using a chromosome 15-specific satellite sequence, D15Z1 (also referred to as satellite III). Cross-hybridization between chromosomes 14 and 15 occurs approximately 12% of the time (Fig. 31.1) [4–6]. A case in which both chromosomes 14 homologs showed a signal with D15Z1 was reported [5]. The chance of both parents carrying the variant 14 is approximately 1 per 69 couples. If both parents are carriers, the chance of the child being homozygous for the variant is 25%. However, if only one parent is a carrier or is available, the possibility of uniparental disomy

H. E. Wyandt, V. S. Tonk, *Human Chromosome Variation: Heteromorphism and Polymorphism*, DOI 10.1007/978-94-007-0896-9_19,

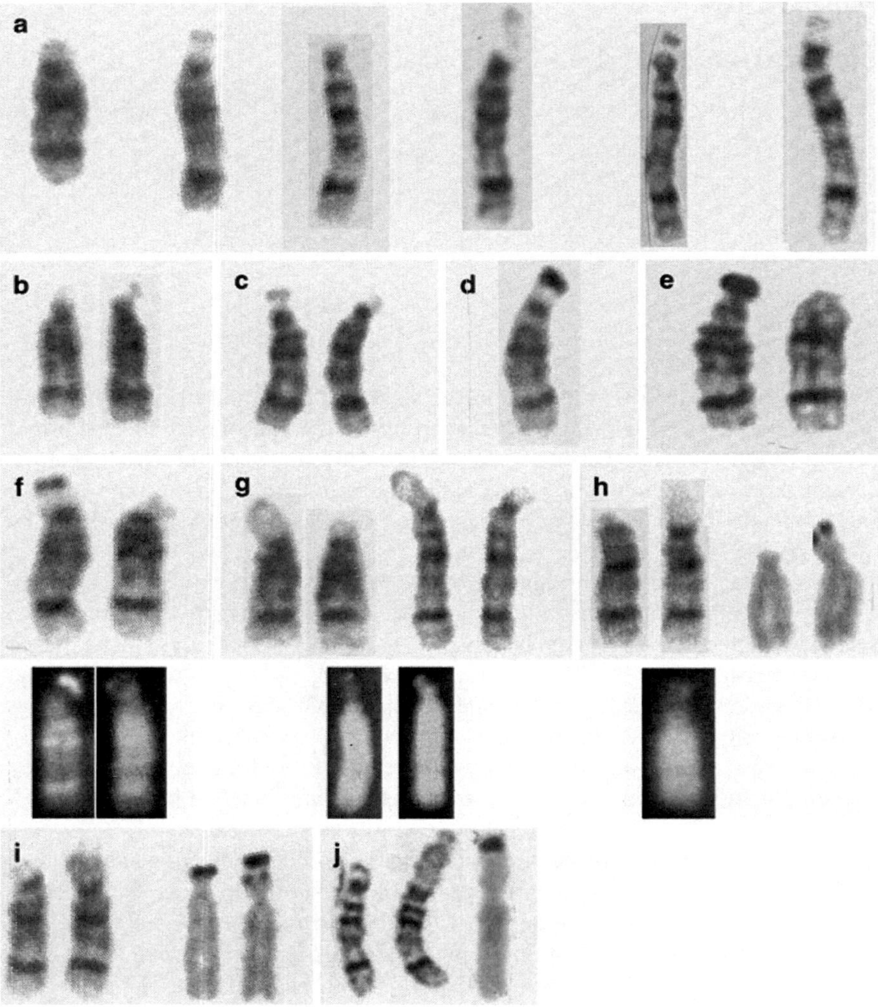

Fig. 19.1 **a** Normal chromosomes 14 from six different subjects by G-banding showing increasing band resolution. **b–c** Pairs of homologs at standard band resolution from two different subjects. **d** Chromosome 14 with moderate-size satellite. **e** Chromosomes 14 from two different people showing large satellite (*left*) and lacking a visible short arm (*right*). **f–h** Pairs of chromosomes by G-banding (*top*) and Q-banding (*bottom*). **h–j** Pairs of chromosomes by G-banding (*left*) with chromosomes from same subjects stained for NOR regions by silver staining (*right*). A double NOR is shown in one of the chromosomes **i** A particularly striking variant of 14p by G-banding (**j**, middle) has a terminal constriction, corresponding to the large NOR by silver staining (**j**, right). [Contributors: **a–d**, **f–j** Center for Human Genetics, Boston University (c1); **e** Jacqueline Schoumans, University Hospital Haukeland (c27, c28)]

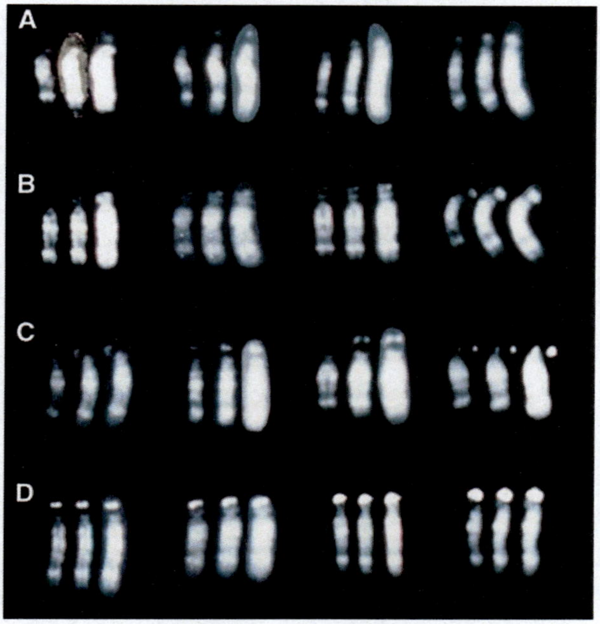

Fig. 19.2 Sixteen different variants of chromosome 14 from a population of 39 unrelated people. Each chromosome is serially printed to reveal heteromorphisms not visible at an exposure generally chosen to define the overall banding pattern [Reproduced from Olson SB et al. (1986). Am J Hum Genet 38:235–252]

for chromosome 14 should be considered, particularly if the proband has clinical abnormalities.

Complete loss of satellite III sequences from 14p and the centromeric region in a case of 14p- has been reported [7]. This presumably normal variant showed absence of the short arm, NOR and possibly part of the centromeric region. A normal alpha satellite signal (D15Z) was obtained.

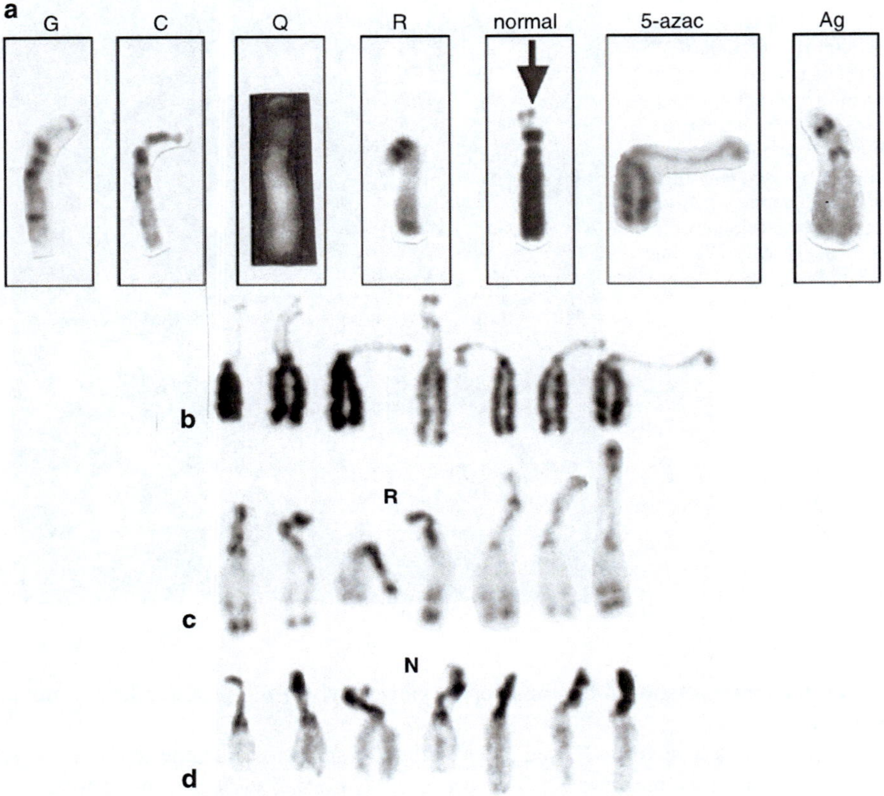

Fig. 19.3 **a** Chromosome 14pss by different banding techniques (G, C, Q, R) and after treatment with 5 azacytidine (5-azacytidine, $5{\times}10^{-6}$ M, last 7 hr of culture). Silver staining at far right (Ag) shows two NOR regions. Unbanded (*normal*) 14pss (*arrow, third from right*) points to untreated chromosome. **b** The same unbanded 14pss chromosome from several mitoses treated with 5-azacytidine, stained with Giemsa. **c** Several chromosomes by RHG-banding. **d** Several chromosomes from same sample by silver staining showing variability in number, size and location of active NOR regions. See Fig. 19.4 and summary for more details. [Contributor: Petr Balicek, Division of Medical Genetics, University Hospital, Hradec Kralove (c16)]

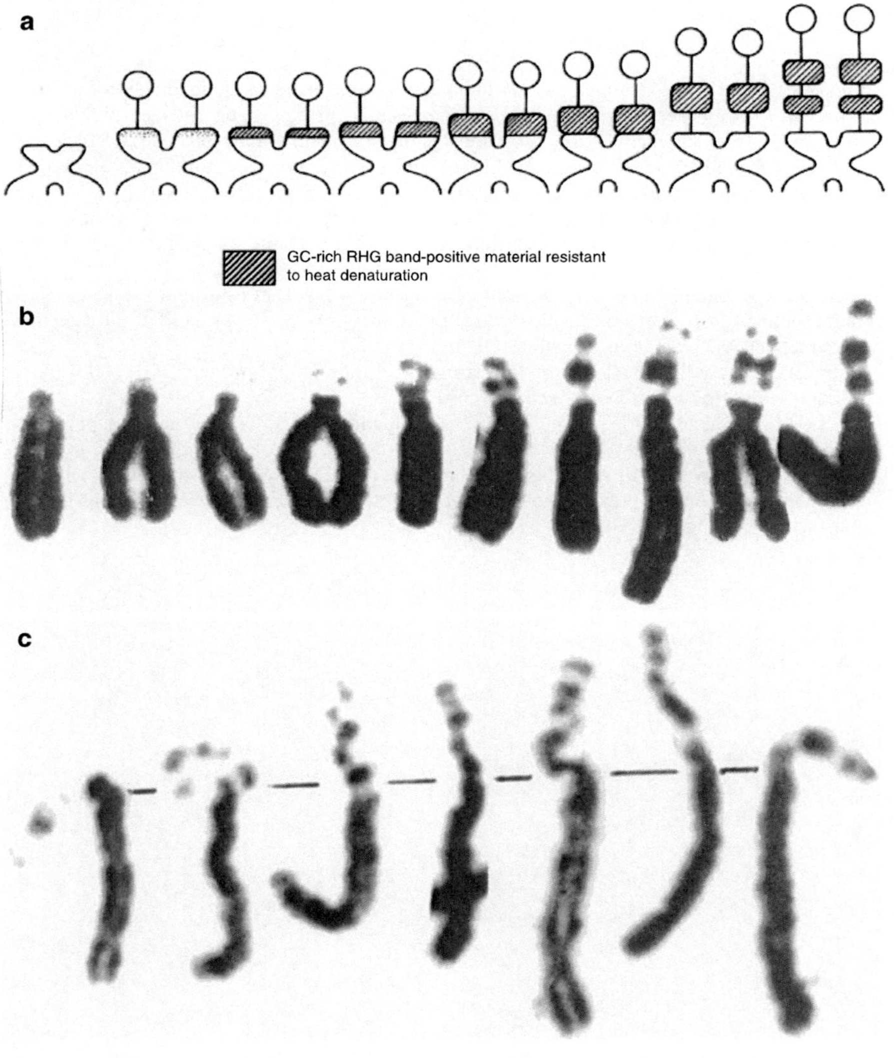

Fig. 19.4 **a** Schematic representation of variability in the amount of GC-rich (RGH-positive) material. Any larger accumulation of such material tends to dissociate from the basal segment of short arms by a proximal secondary constriction. **b** Conventionally stained acrocentric chromosomes with different short arms derived from various individuals. The last chromosomes display double and triple satellites. **c** Chromosome with 14pss from several mitoses in prophase. The intercalar segment of double satellites can be dissociated by secondary constrictions (active NORs) in practically any site of the segment. See summary for more details. [Contributor: Petr Balicek, Division of Medical Genetics, University Hospital, Hradec Kralove (c16)]

References

1. Olson SB, Magenis RE, Lovrien EW (1986) Human chromosome variation: the discriminatory power of Q-band heteromorphism (variant) analysis in distinguishing between individuals, with specific application to cases of questionable paternity. Am J Hum Genet 38:235–252
2. Balicek P, Zizka J (1980) Intercalar satellites of human acrocentric chromosomes as a cytological manifestation of polymorphisms in GC-rich material. Hum Genet 54:343–347
3. Balicek P, Zizka J, Skalska H (1982) RGH-band polymorphism of the short arms of human acrocentric chromosomes and relationship of variants to satellite association. Hum Genet 62:237–239
4. Stergianou K, Gould CP, Waters JJ, Hulten MA (1993) A DA/DAPI positive 14p heteromorphism defined by *in situ* hybridization using chromosome 15-specific probes D15Z1 (Satellite III) and p-TRA-25 (alphoid). Hereditas 119:105–110
5. Shim SH, Pan A, Huang XL, Tonk VS, Varma SK, Milunsky JM, Wyandt HE (2003) FISH variants with D15Z1. Am J Gen Tech 29:146–151
6. Cockwell AE, Jacobs PA, Crolla JA (2007) Distribution of D15Z1 copy number polymorphism. Europ J Hum Genet 15:441–445
7. Earle E, Voullaire LE, Hills L, Slater H Choo KH. (1992) Absence of satellite III DNA in the centromere and the proximal long-arm region of human chromosome 14: analysis of a 14p- variant. Cytogenet Cell Genet 61:78–80

Chapter 20
Chromosome 15

As with the other acrocentric chromosomes, the centromere, short arm and satellite regions of chromosome 15 are variable in size and staining properties (Figs. 20.1–20.3). Current FISH analyses distinguish at least two classes of repetitive DNA in the heterochromatin of chromosome 15: (1) a chromosome 15-specific alpha satellite DNA in the pericentromeric region of chromosome 15 [1] and (2) a chromosome 15-specific subcomponent of satellite III DNA that comprises a major component of the short arm of chromosome 15 [2, 3]. Cross hybridization of a chromosome 15-specific satellite III with other D-group chromosomes was reported in 5.5% of subjects in one study, and was correlated with additional DA/DAPI-positive signals on those chromosomes [4]. Cross hybridization of D15Z1 with chromosome 14 occurs in about 12% of cases (Fig. 31.1) [5–7].

Variations in the size of the classical satellite region of chromosome 15 are also evident by FISH (Fig. 20.4). We know of at least one case with a diminutive short arm on a chromosome 15 by G-banding that shows loss of the classical satellite sequence D15Z1 but has a normal-size alpha satellite signal (Fig. 20.4a, b). In another case, a very large short arm on a chromosome 15 is shown by FISH to have a duplication of the classical satellite sequences (Fig. 20.4d–g). Homology between various satellite III DNA families among acrocentric chromosomes possibly accounts for the fact that almost all breakpoints in Robertsonian translocations occur in the satellite III DNA's [8, 9].

Q-bright satellites (Fig. 20.2), similar in intensity to the Q-bright Yqh region, occur on all of the acrocentric chromosomes with the majority not involving the Y chromosome. Occasionally, however, translocations occur between an acrocentric short arm and the distal long arm of the Y. When they occur, they almost always involve chromosomes 15 or 22. Distinction between Y/15 and Y/22 translocations from other bright variants can be made by staining with DA/DAPI, special treatments such as addition of distamycin A to cultures or, more currently, by FISH analysis (Fig. 20.5 and 20.6).

Also, of interest regarding heteromorphism of chromosome 15, is the close proximity of genes causing Angelman and Prader-Willi syndromes. Zackowski et al. [10] and Butler [11] have questioned whether or not large, dark staining centromeric and short arm variants of chromosome 15 by G-banding (GTG banding) are more

H. E. Wyandt, V. S. Tonk, *Human Chromosome Variation: Heteromorphism and Polymorphism,* DOI 10.1007/978-94-007-0896-9_20,

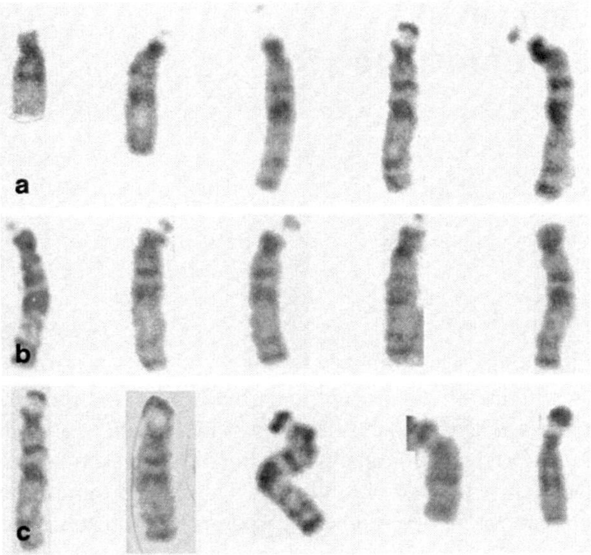

Fig. 20.1 **a** Normal chromosomes 15 from five different subjects by G-banding showing increasing band resolution. **b** Normal chromosomes from 5 subjects showing gradations in size of short arms. **c** Normal chromosomes from 5 different subjects showing variations in size of satellites. [Contributors: **a–c** Center for Human Genetics, Boston University (c1); **b** (*second from right*). Jacqueline Schoumans, Haukeland University Hospital (c19)]

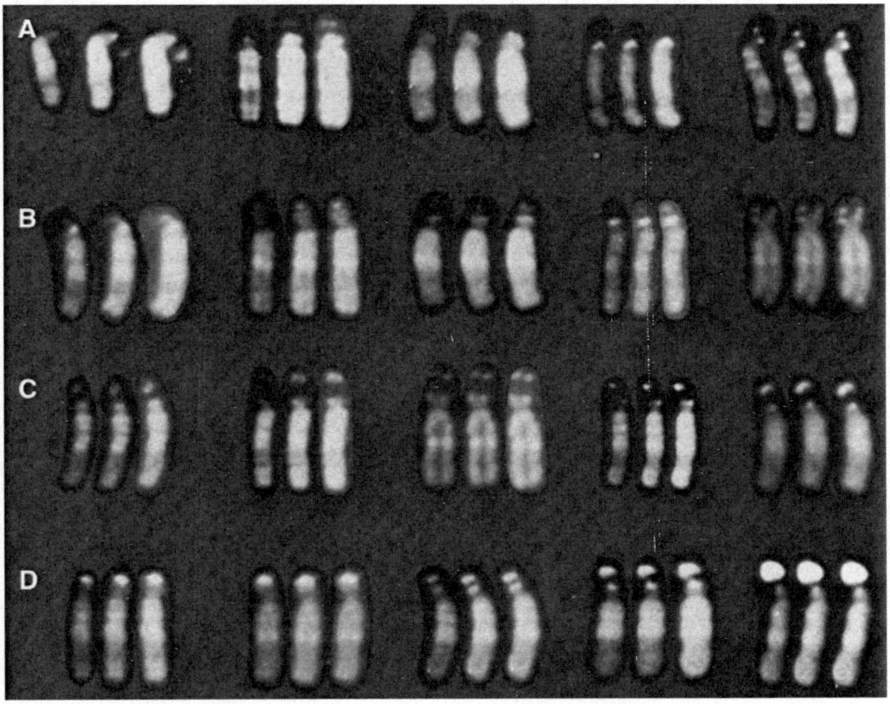

Fig. 20.2 Twenty different variants of chromosome 15 from a population of 39 unrelated people. Each chromosome is serially printed to reveal heteromorphisms not visible at an exposure generally chosen to define the overall banding pattern. The large bright satellites in D5 are relatively uncommon and were present in only one person in the population studied. [Reproduced from Olson SB et al. (1986). Am J Hum Genet 38:235–252]

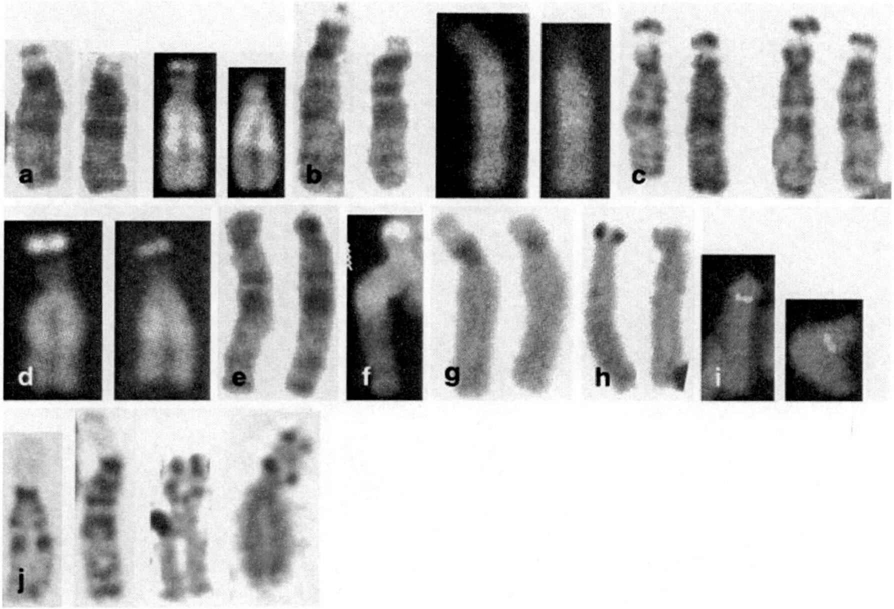

Fig. 20.3 Variant short arm regions by different banding and FISH techniques. **a** Pair of homologs by G and Q-banding showing dark short arm and Q-positive satellite in one homolog (*left*) and narrow Q-positive band in the other (right). **b** Similar comparison in a different subject with a 15p+ (left). Note that the chromosome does not have a Q-bright region, despite size and dark staining of the 15p+. **c–d** Pairs of homologs from a third subject by G-banding (**c**) and by Q-banding (**d**). Note large short arm (**c**, left) and bright satellites on both homologs. **e–i** Note large short arm (**e**, left) has similar G-banding to the 15p+ in (**b**), but has an intensely Q-positive terminal band (**f**) C-banding of same region (**g**, left) is negative and a rather large NOR region is evident by silver staining (**h**, left). Centromeric regions are similar in size in both homologs with the alpha satellite probe, D15Z4 (**i**). **j** Two chromosomes 15 from different mitoses with large pale staining short arm by G-banding (*left*) showing at least two NOR regions by silver staining (two chromosomes at right from different mitoses from same subject). [Contributor: **a–j** Center for Human Genetics, Boston University (c1)]

frequent in the deleted chromosomes than in the normal homologs in Angelman and Prader-Willi syndrome cases. In a FISH study of the deleted chromosomes with D15Z1 (classical satellite probe), Butler points out that variations in the size of the signal were determined to be heritable, but such variations did not correlate with deletion. He concluded that some of the GTG size increases might be due to the deletion bringing non-heterochromatic, GTG-positive regions into close proximity to the centromere. However, the large 15p variants Zackowski et al. observed in AS patients were inherited from the mother.

Euchromatic variants. Rearrangements close to or involving the15q11.2 region are a frequent cause of concern. Deletions of 15q11.2-q13 region are, of course, most often associated with Prader-Willi (PWS) and Angelman (AS) syndromes, identified by loss of the SNRPN probe in PWS and loss of the D15S10 probe in AS. Rare cases of inversion involving the 15q11.2 region occur, and are best studied

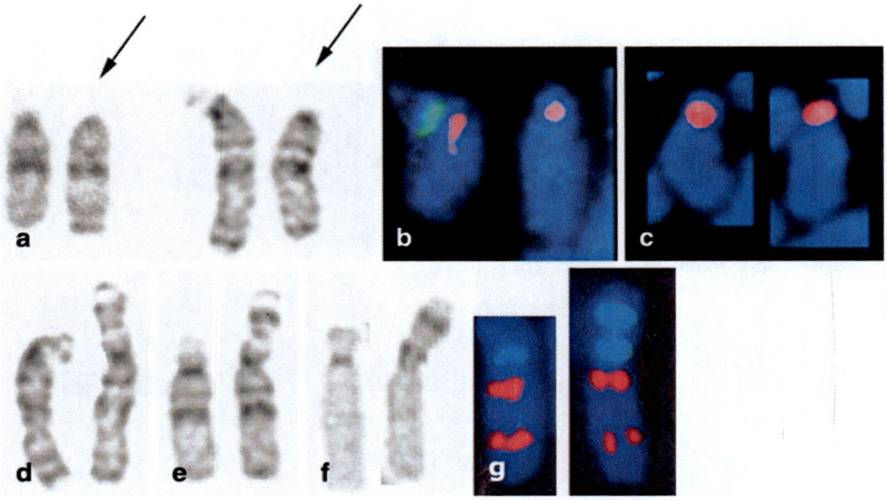

Fig. 20.4 Variants of chromosome 15 studied by FISH. **a–c** Pairs of homolgs showing diminutive short arm (*arrow*). **a** G-banding. **b** FISH, using probe mixture of classical satellite (D15Z1, *green*) and D15S11 (*red*) in band q11.2 showing absence of D15Z1 from one homolog (*right*). **c** Homologs from same subject showing presence of alpha satellite (D15Z4) in centromeric region of both homologs. **d–g** Striking 15p+ from apparently normal subject by G-banding (**d, e**) showing two C-band positive regions (**f**, *right*) and two blocks of classical satellite (D15Z1) by FISH (**g**, *right*). Red signs are for SNRPN (proximal) and PML (distal). [Contributors: **a–c** Center for Human Genetics, Boston University (c1); **d–g** KF Choeng, KK Womens and Childrens Hospital (c31)]

by silver staining. Two cases are shown in Fig. 20.7a, b: one case was previously reported by Jalal and Keterling [12] with no clinical details given, but was presumably clinically normal; the other case, detected prenatally before FISH and molecular techniques were available (Wyandt, unpublished) was terminated. No anomalies were noted and no details were given of the terminated fetus.

Duplication in the 15q11.2 band (Fig. 20.8) is reported in both phenotypically normal and abnormal patients [13, 14]. Ludowese et al. [15] reported ten cases of 15q12.2-q13.1 duplication with apparently normal phenotype. Jalal et al. [16] reported a similar finding in 15 cases from seven unrelated families (Fig. 20.8b, c). Duplications involving the PWS and AS critical regions also result in abnormal phenotypes [17]. Normal patients studied by Riordan and Dawson (1998) [18], using probes for PWS/AS regions, fell into two groups: (1) those with duplication of PWS/AS probes and (2) those without, but who had a large D15Z variant. In general, cases that do not include the PWS/AS region are familial and without phenotypic effect. Cases with developmental delay, minor malformations and/or autism that include the PWS/AS region can be either de novo or familial [19, 20]. Cases with autism or autistic-like behavior are maternally derived whereas paternally derived cases usually have a normal phenotype [19, 21]. The incidence of dup(15) is

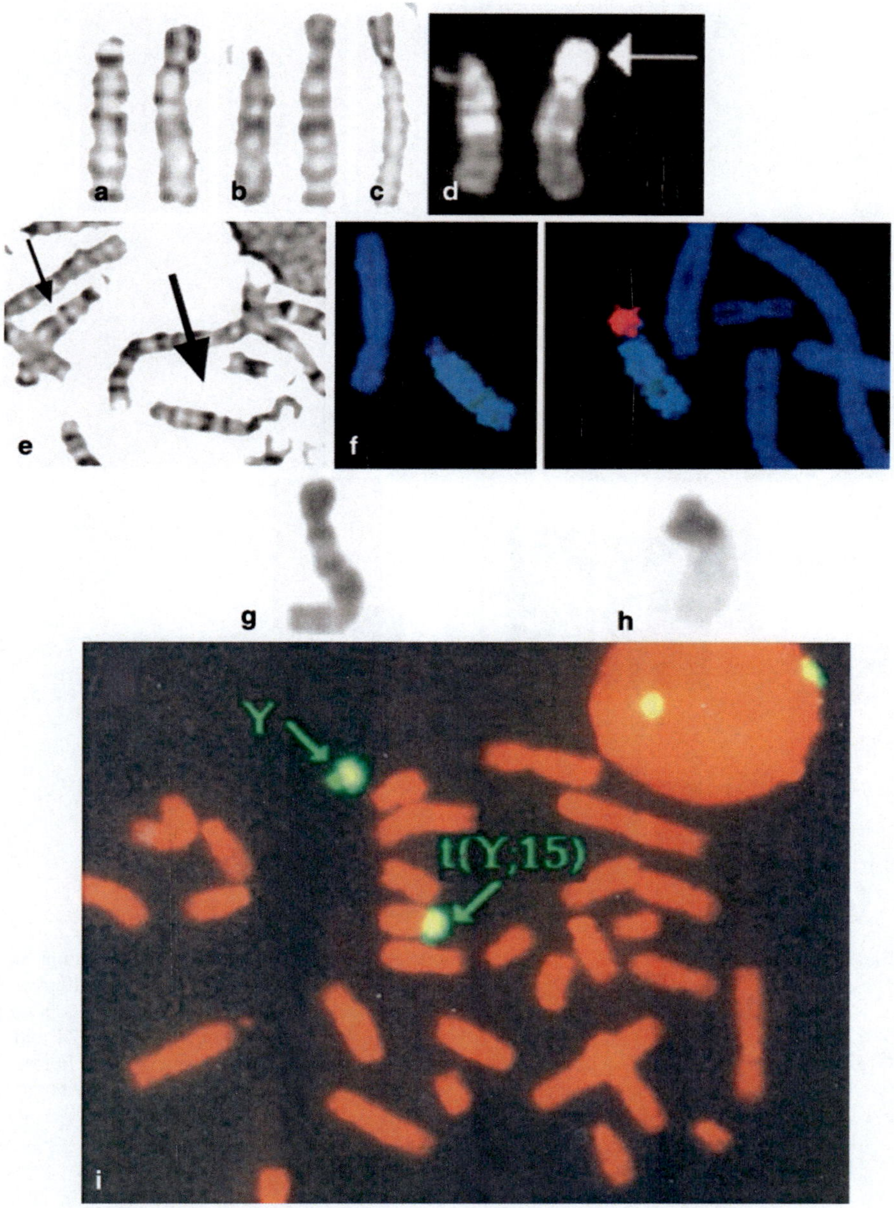

Fig. 20.5 Variants of 15p due to translocations with distal Yq. **a–d** 15p+ by G-banding (**a, b,** *right*), by C-banding (**c**) and by Q-banding (**d**, *right*). **e** Metaphase treated with distamycin A produces under-condensation of 15p+ (*large arrow*) compared to normal homolog (*small arrow*). **f** Fish with painting library for 15 (wcp15, *green*) and distal Yq (DYZ1, *red*). Note absence of red signal from normal homolog (*left*). **g–i** Different normal male subject with 15p+ by G-banding (**g**), C-banding (h) and by FISH with DYZ1 (**i**). Note two green signals in interphase in the same metaphase (*upper right*). [Contributors: **a–f** KF Choeng, KK Womens and Childrens Hospital (c32); **g–i** J M Fink, Hennepin County Medical Center (c37)]

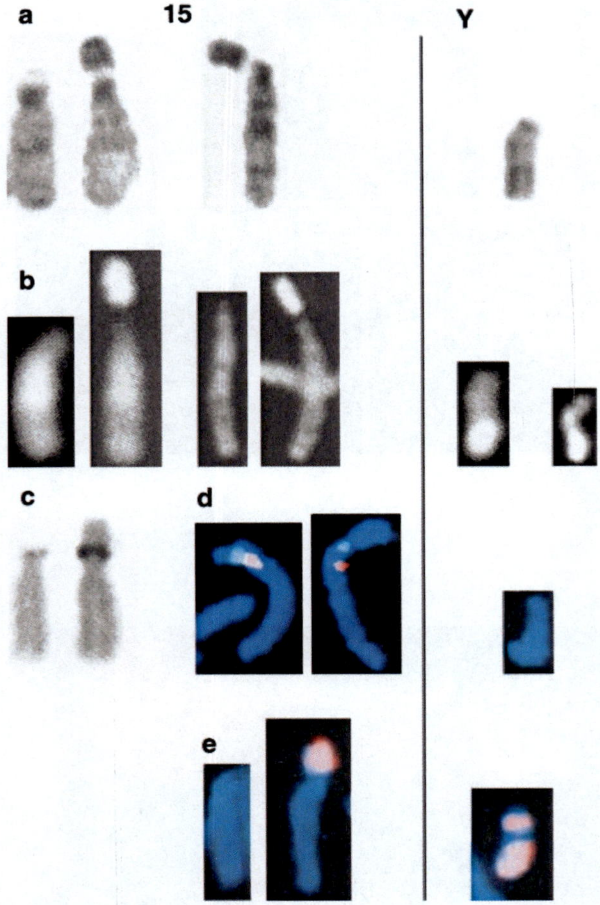

Fig. 20.6 Apparently large satellite on 15p due to translocation from distal Yq. **a** G-banding of 15ps+ from two different cells and of Y from one cell. **b** Q-banding of 15s and Y from two different cells. **c** Silver staining showing large NOR corresponding to constriction in 15p+. Note small NOR in normal homolog (*left*). **d** Chromosomes 15 and Y by FISH with probe mixture of D15Z1 and D15Z4 showing classical and alpha satellites on both homologs. **e** Chromosome painting library (wcpY, including DYZ1) showing signals over both 15 ps+ and the Y chromosome. [Contributor: **a–e** Center for Human Genetics, Boston University (c1)]

unknown, but the estimated prevalence among patients with mental retardation and/or developmental delay is 0.17–0.5% [22]. From these observations it is evident that any suspected case of dup(15)(q11q13) should have parental chromosome studies and be tested with FISH probes to determine whether or not the PWS/AS region is duplicated.

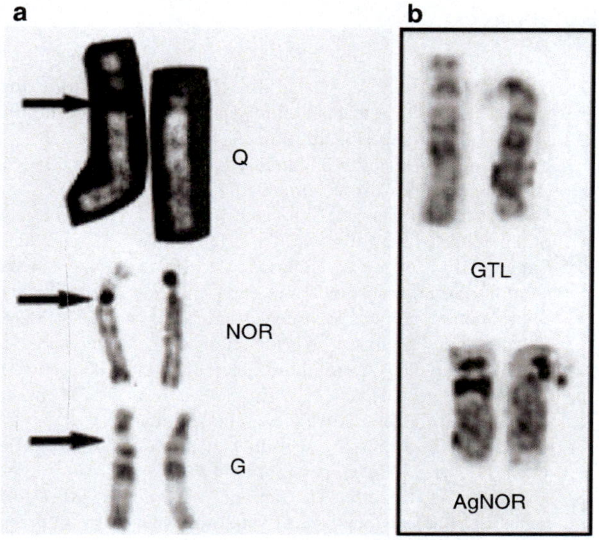

Fig. 20.7 Two *de novo* cases with an unstained area at 15q11.2 that is positive by AgNOR staining indicating an insertional translocation or inversion of the nucleolar organizing region from the acrocentric short arm: **a** (Q-banded, top, G-banded, bottom, Ag-NOR middle) (Wyandt, unpublished). Similar case reported by Jalal and Ketterling [12]. [Reproduced from Plate 47A, Atlas of Human Chromosome Heteromorphisms (Wyandt HE Tonk V, eds), Kluwer, Dordrect, 2004] **b** G-banded (*top*) and Ag NOR stained (*bottom*)

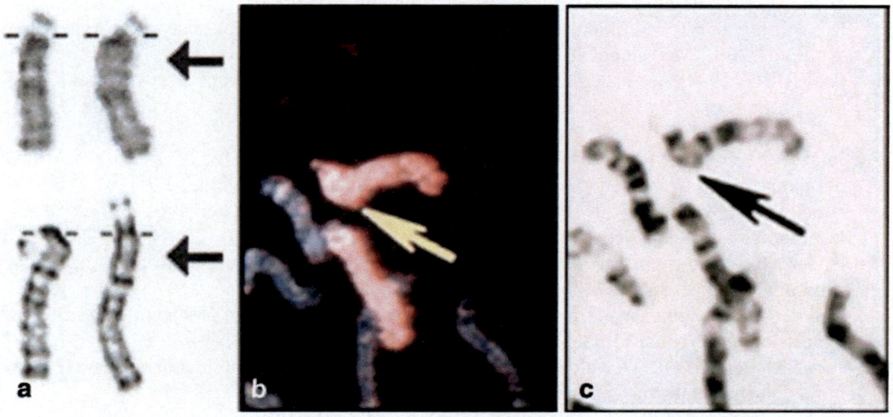

Fig. 20.8 **a** Pairs of 15s from a normal subject showing prominent euchromatic band 15q11.2. **b, c** Duplication variant of 15q11.2-q13.1 in a sequential G-banded partial metaphase (**b**), followed by wcp15 probe analysis (**c**), reported by Jalal et al. [16] in 15 cases from 7 unrelated families. [Reproduced from Plate 47 D and E, Atlas of Human Chromosome Heteromorphisms (Wyandt HE Tonk V, eds), Kluwer, Dordrect, 2004. Contributor: **a** Arturo Anguiano, Quest Diagnostics (c29)]

References

1. Choo KH, Earle E, Vissel B Filby RG (1990) Identification of two distinct subfamilies of alpha satellite DNA that are highly specific for human chromosome 15. Genomics 7:143–151
2. Higgins MJ, Wang HS, Shtromas I, Haliotis T, Roder JC, Holden JJ White BN (1985) Organization of a repetitive human 1.8 kb KpnI sequence localized in the heterochromatin of chromosome 15. Chromosoma 93:77–86
3. Fowler C, Drinkwater R, Skinner J Burgoyne L (1988) Human satellite–III DNA: an example of a macrosatellite polymorphism. Hum Genet 79:265–272
4. Smeets DF, Merkx GF, Hopman AH (1991) Frequent occurrence of translocations of the short arm of chromosome 15 to other D-group chromosomes. Hum Genet 87:45–48
5. Stergianou K, Gould CP, Waters JJ, Hulten MA (1993) A DA/DAPI positive 14p heteromorphism defined by *in situ* hybridization using chromosome 15-specific probes D15Z1 (Satellite III) and p-TRA-25 (alphoid). Hereditas 119:105–110
6. Shim SH, Pan A, Huang XL, Tonk VS, Varma SK, Milunsky JM, Wyandt HE (2003) FISH variants with D15Z1. Am J Gen Tech 29:146–151
7. Cockwell AE, Jacobs PA, Crolla JA (2007) Distribution of D15Z1 copy number polymorphism. Europ J Hum Genet 15:441–445
8. Gravholt CH, Friedrich U, Caprani M, Jorgensen AL (1992) Breakpoints in Robertsonian translocations are localized to satellite III DNA by fluorescence *in situ* hybridization. Genomics 14:924–930
9. Bandyopadhyay R, Berend SA, Page SL, Choo KH Shaffer LG (2001) Satellite III sequences on 14p and their relevanceto Robertsonian translocation formation. Chromosome Res 9:235–242
10. Zackowski JL, Nicholls RD, Gray BA, Bent-Williams A, Gottlieb W, Harris PJ, Waters MF, Driscoll DJ, Zori RT, Williams CA (1993) Cytogenetic and molecular analysis in Angelman syndrome. Am J Med Genet 46:7–11
11. Butler MG (1994) Are specific short arm variants or heteromorphisms over-represented in the chromosome 15 deletion in Angelman or Prader-Willi syndrome patients? Am J Med Genet 50:42–45
12. Jalal SM, Ketterling RP (2004) Euchromatic Variants. In: Wyandt HE, Tonk VS (eds) Atlas of human chromosome heteromorphisms. Kluwer Academic, Dordrecht, pp 75–95
13. Brookwell R, Veleba A (1994) Proximal 15q variant with normal phenotype in three unrelated individuals. Clin Genet 31:311–314
14. Butler MG, Greenstein MA (1991) Molecular cytogenetics of Prader-Willi and Angelman syndromes. Lancet 338:1276
15. Ludowese CJ, Thompson KJ, Sekon GS, Pauli RM (1991) Absence of predictable phenotypic expression in proximal 15q duplications. Clin Genet 40:194–201
16. Jalal SM, Persons OL, Dewald GW (1994) Form of 15q proximal duplication appears to be a normal euchromatic variant. Am J Med Genet 52:495–497
17. Mao R, Jalal SM, Snow K, Michels VV, Szabo SM, Babovic-Vuksanovic D (2000) Characteristics of two cases with dup(15)(q11.2-12): one of maternal and one of paternal origin. Genet In Med 2:131–135
18. Riordan D, Dawson AJ (1998) The evaluation of proximal 15q duplications by FISH. Clin Genet 54:517-521
19. Browne CE, Dennis NR, Maher E et al (1997) Inherited interstitial duplications of proximal 15q: genotype-phenotype correlations. Am J Med Genet 61:1342–1352
20. Bolton PF, Dennis NR, Browne CE, Thomas NS, Veltman MWM, Thompson RJ, Jacobs P (2001) The phenotypic manifestations of interstitial duplications of proximal 15q with special reference to the autistic spectrum disorders. Am J Med Genet (Neuro Genet) 105:675–685
21. Yardin C, Esclaire F, Laroche C, Terro F, Barthe D, Bonnefont J-P, Gilbert B (2002) Should the chromosome region 15q11q13 be systematically tested by FISH in the case of an autistic-like syndrome? Clin Genet 61:310–313
22. Moeschler JB, Mohandas TK, Hawk AB, Knoll W (2002) Research Letter. Estimate of the prevalence of proximal 15q duplication syndrome. Am J Med Genet 111:440–442

Chapter 21
Chromosome 16

Approximately 38% of individuals, in the New Haven study of 7- and 8-year olds [1], showed a level-1 size 16qh region. Two to 4% showed level-5 size variants. In another study of 35 unrelated subjects studied by C-banding, 11 variants of chromosome 16 were found, nine of which were level 1 [2]. No cases of 16qh inversion were reported in these studies or in a study by Hsu et al. [3] of 6,250 prenatal specimens representing four major population groups. Potluri et al. [4] also found no inversion 16 in 200 New Delhi infants. Figure 21.1 shows size variations of 16qh region in a typical laboratory sample of selected normal subjects ranging from level 1 to far exceeding level 5.

More recent molecular techniques have shown additional variants. A rare familial variant consisting of a C-negative band inserted in the center of a large 16qh region, was shown to hybridize with 16-specific alphoid sequences, interpreted by Jalal et al. [5] to be due to an inverted duplication that included the centromeric sequences. A very small 16qh region that was C-negative showed a normal centromeric signal by *in situ* hybridization [6].

As with the other major heteromorphic regions, numerous studies have attempted to associate striking heteromorphisms of chromosome 16qh with deleterious effects. Soudek et al. [7] found a higher frequency of 16qh- in retarded individuals than in a control group and Buretic-Tomljanovich et al. [8] suggested an increase in heterochromatin of chromosome 16 in couples with a stillborn or a malformed child. Variations in size and location of the major heterochromatic regions have particularly been implicated in various cancers and leukemias (Table 4.1).

Euchromatic variants: Euchromatic variants of chromosome 16 are reported for both the long and short arms (Fig. 21.2). Reports of deletion of 16q21 have been contradictory [9]. A *de novo* deletion of 16q13-q22 was reported in a one-year-old girl with growth retardation and multiple dysmorphic features [10]. However, deletion of euchromatic band 16q21 in phenotypically normal individuals, is described in several reports [11, 12], including one report of the deletion in a mother, son, and newborn daughter [13]. Callen et al. [12], using somatic cell hybrids and FISH analysis, confirmed the loss of 16q21 in a phenotypically normal patient, and observed that reported deletions in patients with phenotypic anomalies involve larger segments extending proximal or distal to 16q21.

H. E. Wyandt, V. S. Tonk, *Human Chromosome Variation: Heteromorphism and Polymorphism,* DOI 10.1007/978-94-007-0896-9_21,

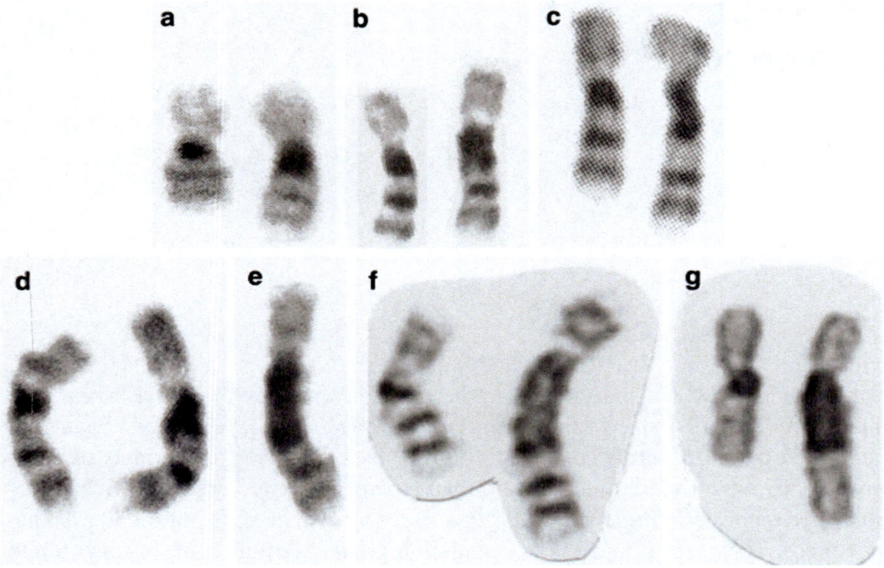

Fig. 21.1 **a** Normal chromosomes 16 from five different subjects by G-banding showing increasing band resolution. **b–e** Pairs of chromosomes 16 from four different subjects showing 16qh+ (*right hand* chromosome in each pair). **f** Chromosome 16 by G-banding with extremely long qh region. **g** Pair of chromosome 16 homologs by G-banding showing striking discrepancy between a short qh region and a very long qh region in the same person. Pair of C-banded chromosomes from the same subject (*extreme right*). [Contributors: **a–e** Center for Human Genetics, Boston University (c1); f Jacqueline Schoumans, Haukeland University Hospital (c21); **g** Peter Benn, University of Connecticut Health Center (c11)]

An extra C-band-negative segment in the proximal short arm of chromosome 16 that was late replicating and apparently innocuous was described by Thompson and Roberts [14, 15]. Additional cases are reported by others [16–19]. In a report by Jalal et al. [17], 16p arms that were about one-third larger than normal were present in two unrelated infants with different anomalies. The variant was present in the normal father and normal paternal grandmother in one case and in a mother with minor anomalies in the other. Bryke et al. [18], reporting a similar finding in a fetus and normal mother, showed the C-band- negative region was also negative for α-satellite DNA.

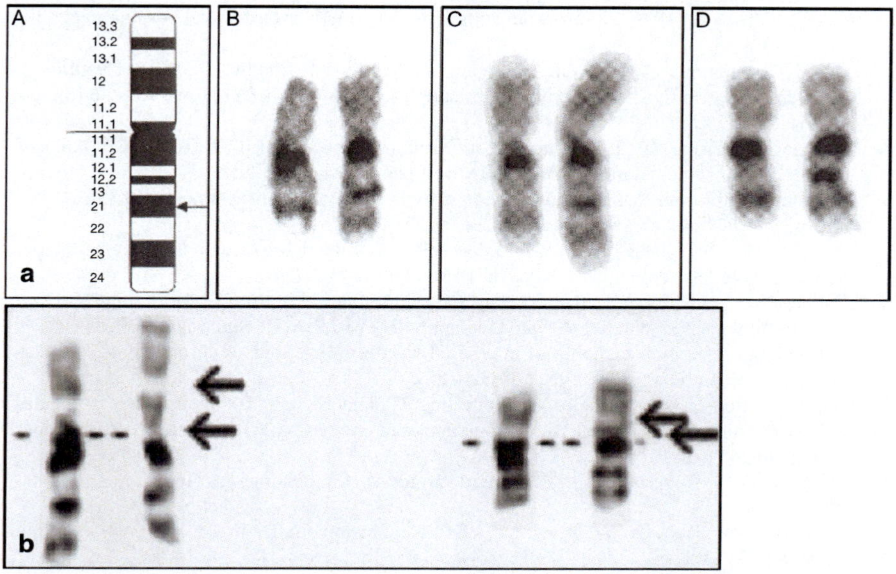

Fig. 21.2 Euchromatic variants in chromosome *16*. **a** Chromosome 16 ideogram and the homologous pairs from the fetus, mother, and a male sibling. (*A*) Ideogram at 400 band stage (*arrow* indicates the 16q21 band). Pair 16 from the fetus (*B*), the mother (C) and a male sibling (*D*) with the deleted 16q21 band in each pair (chromosomes at *left*). **b** Increase of 16p arm by about one-third as a duplication variant. [Reported by Jalal et al. [17]; reproduced from Plate 40, Atlas of Human Chromososme Heteromorphisms (Wyandt HE and Tonk V, eds), Kluwer, Dordrecht, 2004]. This has been reported in a number of cases [14–19]

References

1. Lubs HA, Patil SR, Kimberling WJ, Brown J, Cohen M, Gerald P, Hecht F, Myrianthopoulos N, Summit RL (1977) Q and C-banding polymorphisms in 7 and 8 year old children: racial differences and clinical significance. In: Hook EB, Porter IH (eds) Population cytogenetic studies in humans. Academic Press, New York, pp 133–159
2. Craig-Holmes AP (1977) C-band polymorphisms in human populations. In: Hook EB, Porter IH (eds) Population cytogenetics. Studies in humans. Academic Press, New York, pp 161–177
3. Hsu LYF, Benn PA, Tannenbaum HL, Perlis TE, Carlson AD (1987) Chromosomal polymorphisms of 1,9, 16 and Y in 4 major ethnic groups: a large prenatal study. Am J Med Genet 26:95–101
4. Potluri VR, Singh IP, Bhasin MK (1985) Chromosomal heteromorphisms in Delhi infants. III. Qualitative analysis of C-band inversion heteromorphisms of chromosomes 1, 9 and 16. J Heredity 76:55–58
5. Jalal SM, Law ME, DeWald GW (1993) Inverted duplication involving alpha satellite DNA resulting in a C-negative band in the qh region of chromosome 16. Am J Med Genet 46:351–352
6. Verma RS, Luke S, Mathews T Conte RA (1992) Molecular characterization of the smallest secondary constriction region (qh) of human chromosome 16. Genet Anal Tech Appl 9:140–142

7. Soudek D, Sroka H (1979) Chromosomal variants in mentally retarded and normal men. Clin Genet 16:109–116
8. Buretic-Tomljanovic A, Rodojcic Badovinac A, Vlastelic I, Randic LJ (1997) Quantiative analysis of constitutive heterochromatin in couples with fetal wastage. Am J Reprod Immun 38:201–204
9. Jalal SM, Ketterling RP (2004) Euchromatic variants. In: Wyandt HE, Tonk VS (eds) Atlas of human chromosome heteromorphisms. Kluwer, Dordrecht, pp 75–95
10. Naritomi K, Shiroma N, Izumikawa Y, Sameshima K, Ohdo S, Hirayama K (1988) 16q21 is critical for 16q deletion syndrome. Clin Genet 33:372–375
11. Witt DR, Lew SP, Mann J (1988) Heritable deletion of band 16q21 with normal phenotype: relationship to late replicating DNA. Am J Hum Genet 43:A127
12. Callen DF, Eyre H, Lane S, Shen Y, Hansmann I, Spinner N, Zackai E, McDonald-McGinn D, Schuffenhauer S, Wauters J, Van Thienen M-N, Van Roy B, Sutherland GR, Haan EA (1993) High resolution mapping of interstitial long arm deletions of chromosome 16: relationship to phenotype. J Med Genet 30:828–832
13. Hand JL, Michels VV, Marinello MJ, Ketterling RP, Jalal SM (2000) Inherited interstitial deletion of chromosomes 5p and 16q without apparent phenotypic effect: further confirmation. Prenat Diagn 20:144–148
14. Thompson PW, Roberts SH (1987) A new variant of chromosome 16. Hum Genet 76:100–101
15. Thompson PW, Roberts SH, Rees SM (1990) Replication studies in the 16p+ variant. Hum Genet 84:371–372
16. Pinel I, de Bustamante AD, Urioste M, Felix V, Ureta A, Martinez-Frias ML (1988). An unusual variant of chromosome 16. Two new cases. Hum Genet 80:194
17. Jalal SM, Schneider NR, Kukolich MK, Wilson GN (1990) Euchromatic 16p+ heteromorphism: first report in North America. Am J Med Genet 37:548–550
18. Bryke CR, Breg WR, Potluri VR, Yang-Feng TL (1990) Duplication of euchromatin without phenotypic effects: a variant of chromosome 16. Am J Med Genet 36:43–44
19. Verma RS, Kleyman SM, Conte RA (1997) Variant euchromatic band within 16q12.1. Clin Genet 52:446–447

Chapter 22
Chromosome 17

Large (Fig. 22.1a) and small variants of the centromeric region on chromosome 17 were reported in the New Haven study of 7- and 8-year-olds [1] with a combined frequency of about 6%. A small centromeric region was more frequent in white children and the larger region was more frequent in black children. In an Estonian population of normal adults, Mikelsaar et al. [2] reported 17ph+ to be more frequent in men than women.

Satellited 17: Several investigators [3–5] reported cases with an apparent secondary constriction in the short arm of chromosome 17. Kuleshov and Kulieva [6], in a study of 6,000 newborns, also mention a case with satellites on the short arms of 17 or 18. The constriction in these studies was not in every cell and was often present in low frequency. The variant was initially ascertained in subjects with various phenotypic abnormalities including trisomy 21, congenital heart disease, and cri-du-chat syndrome. In several instances normal carriers were reported as having aneuploid offspring [7, 8], and the variant was observed in couples having multiple abortions [9]. Despite these observations, the marker was also found in normal individuals [10], or was often familial and carried by normal relatives [11, 12]. In at least one family, it could be traced back several generations.

Schmid and Bauchinger [3] described the marker in five subjects with frequencies ranging from 10 to 100%. The frequency of satellite association was 2%. Others, observing constrictions in 17p (4, 5, 13), found no evidence of rearrangement or association with D or G group chromosomes. Oliver et al. [13] studied four patients with 17p variants by silver staining and found no NOR-positive chromosomes 17. Similarly, Patil and Bent [9] compared staining and satellite association of a 17p variant with D and G group chromosomes, and found no silver staining or satellite association of the 17p variant.

Fragile site at 17p12: Some cases of reported satellited 17 may in fact represent a rare, heritable, fragile site on chromosome 17, fra(17)(p12) [14–16]. At least one case of homozygosity for this site has been reported in a healthy man [16]. Other studies [17] have shown a higher than expected incidence of fra(17)(p12) in patients with hematological disorders.

Identification of satellited 17: Differentiation of a satellited 17 from a fragile site is definitive when results are positive with Ag-NOR staining (Fig. 22.1b). Nega-

H. E. Wyandt, V. S. Tonk, *Human Chromosome Variation: Heteromorphism and Polymorphism,* DOI 10.1007/978-94-007-0896-9_22,

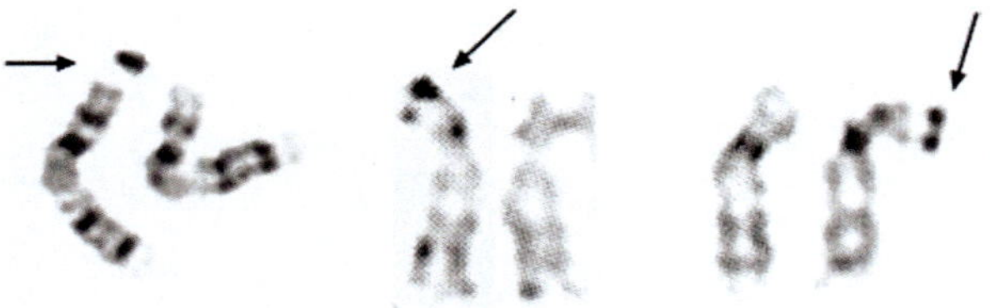

Fig. 22.1 Satellited 17p region **a** by G-banding and **b, c** by silver staining. *Arrows* point to NOR region,visible as a *constriction* by G-banding and as *dark staining* by silver staining. [Contributor, James M. Fink, Hennepin County Medical Center (c37)].

tive results may require chromosome painting to identify the material distal to the constriction as being from chromosome 17. In cases with a fragile site, all cells do not usually show the site. If heritable, the frequency of the site may be increased by special induction [18].

References

1. Lubs HA, Patil SR, Kimberling WJ, Brown J, Cohen M, Gerald P, Hecht F, Myrianthopoulos N, Summit RL (1977) Q and C-banding polymorphisms in 7 and 8 year old children: racial differences and clinical significance. In: Hook EB, Porter IH (eds) Population cytogenetic studies in humans. Academic Press, New York, pp 133–159
2. Mikelsaar A-V, Tuur SJ, Kaosaar ME (1973) Human karyotype polymorphism. I. Routine and fluorescence microscopic investigation of chromosomes in a normal adult population. Humangenetik 20:89–101
3. Schmid E, Bauchinger M (1969) Structural polymorphism in chromosome 17. Nature 221:387–388
4. Ferguson-Smith MA, Ferguson-Smith ME, Ellis PM, Dickson M (1962) The sites and relative frequencies of secondary constrictions in human somatic chromosomes. Cytogenetics 1:325–43
5. Moores EC, Anders JM, Emanuel R (1966) Inheritance of marker chromosomes from a cytogenetic survey of congenital heart disease. Ann hum Genet 30:77–84
6. Kuleshov NP, Kulieva LM (1979) Chastota khromosmnykh variantov v populiatsiiakh cheloveka (Frequency of chromosome variants in human populations). Genetika 15:745–751
7. Berg JM, Fauch JA, Pendrey MJ, Penrose LS, Ridler MA, Shapiro A (1969) A homozygous chromosomal variant. Lancet 1(7531):531
8. Gustaavson KH, Kjessler B (1978) A variant chromosome 17 in a mother with repeated abortions and 46,XY/47,XXY Klinefelter son. Upsala J med Sci 83:119–122
9. Patil SR, Bent FC (1980). Silver staining and the 17ps chromosome. Clin Genet 17:281–284
10. Verma RS, Ved BS, Warman J, Dosik H (1979) Clinical significance of the satellited short arm of human chromosome 17 (17ps+): a rare heteromorphism? Ann Genet 22:133–136
11. Priest JH, Peakman DC, Patil SR, Robinson A (1970) Significance of 17ps+ in three generations of a family. J Med Genet 7:142–147
12. Sandstrom M, Jenkins EC (1973) A 17p marker chromosome familial study. Ann Genet 16:267–269
13. Oliver N, Francke U, Taylor KM (1978) Silver staining studies on the short arm variant of human chromosome 17. Hum Genet 42:79–82

14. Sutherland GR (1979) Heritable fragile sites on human chromosomes. I. Factors affecting expression in lymphocyte culture. Am J Hum Genet 31:125–135
15. Shabtai F, Klar D, Halbrecht I (1982). Chromosome 17 has a real fragile site at p12. Hum Genet 61:177–179.
16. Izakovic V (1984) Homozygosity for fragile site at 17p12 in a 28-year old healthy man. HumGenet 68:340–341
17. Murata M, Takahashi E, Minamihisamatsu M, Ishihara T, Wong P, Bessho M, Hirashima K, Hori T (1988) Heritable rare fragile sites in patients with leukemia and other hematologic disorders. Cancer Genet Cytogenet 31:95–103
18. Schmid M, Feichtinger W, Deubelbeiss C, Weller E (1987) The fragile site (17)(p12): induction by AT-specific DNA ligands and population cytogenetics. Hum Genet 77:118–121

Chapter 23
Chromosome 18

In an early C-banding study of 20 unrelated individuals, Craig-Holmes and Shaw [1] found a small centromeric region (cen-) on an E-group chromosome in one individual and no cases with cen+. By Q and C banding, McKenzie and Lubs [2] found variations in size of the centromeric regions of all of the chromosomes except the Y, including chromosome 18. Müller et al. [3], found chromosome 18 with a double size C-band in about 1.1% of 375 newborns, and in Q- and C-band studies of 7- and 8-year olds, Lubs et al. [4] found a large 18cen in about 8% of white children and in about 15% of black children. Heterochromatin in 18p was found in 1.2% of both races.

Since these early studies, several reports have described heteromorphisms of chromosome 18, usually due to heterochromatin in the short arm (18ph+), usually without phenotypic effect [5–8]. Beaverstock et al. [9] report a chromosome 18 in a Spanish family, with heterochromatin that extended from the centromere into the long arm, lightly stained by G-banding, but C-band positive (Fig. 23.1a). Detected prenatally in both uncultured trophoblasts and cultured villi, the variant was eventually determined to be paternal in origin. Bonfatti and colleagues [8] described pericentric regions of 18 that were nearly undetectable by C-banding, but were detectable with alpha satellite FISH probes. An example is Fig. 23.1b, c.

Babu and Verma [10] studied the heritability of *AluI*-resistant heteromorphisms of pericentric heterochromatin in chromosome 18 and determined the paternal origin of the extra chromosome in a case of trisomy 18 (Fig. 23.2). In a subsequent study of 50 normal Caucasians and five cases of trisomy 18, Babu et al. [11] describe five size-classes, based on comparison with the length of 18p, ranging from negative (class 1) to very large (class 5). The relative incidences of these classes were: (1) 11.3%, (2) 19.1%, (3) 29.57%, (4) 29.57% and (5) 10.4%. Heterochromatin was also classified by location into 4 types: I, absent (11.3%); II, on p-arm only (62.6%); III, on q-arm only (0.87%; and IV, on centromere, extending into both p and q (25.22%). Classes 1 and 5 were predominantly in the trisomy 18 cases.

Euchromatic variants: Wolff et al. [12] reported a prenatal case with duplication of the entire short arm of chromosome 18 that was also present in the phenotypi-

H. E. Wyandt, V. S. Tonk, *Human Chromosome Variation: Heteromorphism and Polymorphism,* DOI 10.1007/978-94-007-0896-9_23,

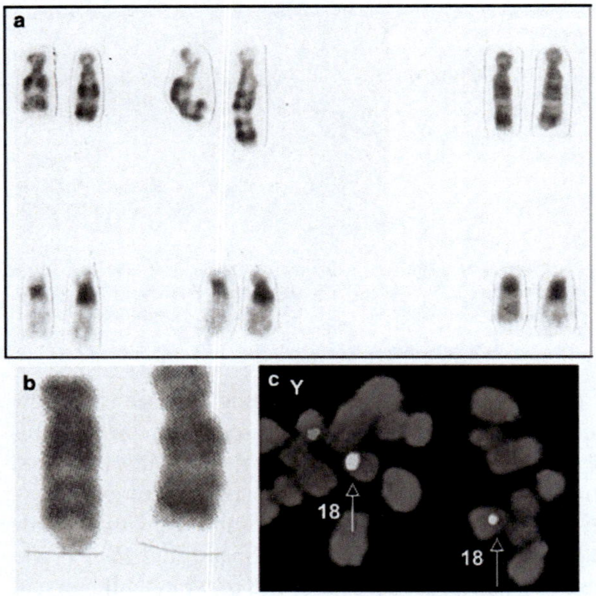

Fig. 23.1 **a** Two chromosome pairs from fetus (*left*) and one pair of chromosomes 18 from the father (*right*). *Top*: GTG-banding: *bottom*: CBG-banding. **b**, **c** Pair of 18s with discrepant size centromeric regions by GTG-banding (**b**) and partial metaphase showing obvious discrepancy in size of alpha satellite signals (CEP18, Vysis, Downers Grove, IL) in the homologs in metaphase (**c**) Both **a** and **b** are from a prenatal sample that showed a low percentage of non-cultured amniocyte cells with two alpha satellite signals for chromosome 18 in by FISH with the same probe. [Contributor: a reproduced with permission from Beaverstock GC et al. (1997). Prenat Diagn 17:585, 586) **b, c** Texas Tech University Health Sciences Center (c43)]

cally normal mother and was confirmed to be of 18p origin by DNA dosage studies (Fig. 23.3). The pregnancy continued, and a normal female was born. According to the authors, some 14 reported cases of 18p duplications result in either a normal phenotype or mild and inconsistent abnormalities. However, other cases are reported with duplication [13, 14] or tetrasomy of 18p that are associated with clinical abnormalities ranging from mild [14, 15] to severe [16, 17]. Causes of the apparent range in phenotypes associated with 18p duplications remain unexplained.

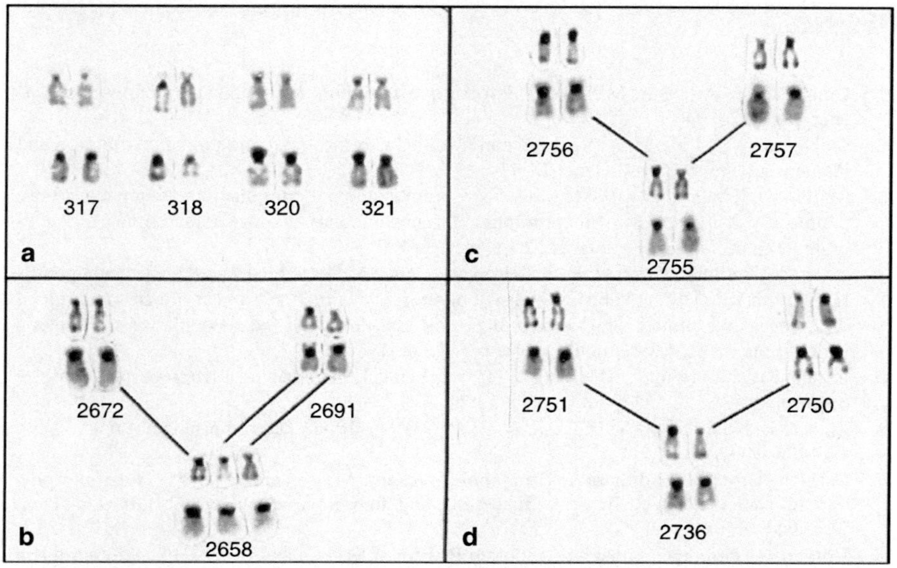

Fig. 23.2 **a** The chromosome 18s from four different individuals treated with *Alu*I and stained by Giemsa (*top row*). The corresponding C-bands are shown below from the same individuals. The *Alu*I-resistant bands represent a small fraction of the C-bands and are usually located distally in the short arm (317, 320 and 321). Its location in the long arm (*left* homolog of 318) and the total absence (*right* homolog of 318) are normal variants. **b** Chromosome 18s from a family with Edward syndrome stained with *Alu*I (*top*) and CBG (*bottom*) techniques. The chromosomes from the mother (2,672) are shown at the *left* while the father's (2,691) are presented at the *right*. One of the chromosomes *18 (far left)* in the proband (2,658) is similar to the pattern seen in the mother while the remaining two (*middle and right*) closely resemble those of the father's, demonstrating the paternal origin. **c** and **d** The chromosome 18s from two normal families stained with *Alu*I (*top*) and CBG (*bottom*) techniques are also included to demonstrate their parental origin. The maternal chromosomes are shown to the *left* (2,756 and 2,751), while the paternal ones are presented to the *right* (2,757 and 2,750). The parental identification is also marked by *lines*. [Reproduced from Babu A and Verma RS (1986) Am J Hum Genet 38:549–554]

Fig. 23.3 Partial karyotype depicting **a** diagrammatic representation of the normal 18 (*left*) and the duplicated 18p. **b** Prometaphase GTG-banded normal 18s (*left in both pairs*) and the duplication 18p chromosomes. *Arrows* denote the duplicated segment. **c** Partial C-banded karyotype revealing normal C-bands on E-group chromosomes including the normal number 18 and the duplicated 18. [Reproduced with permission from Wolff DJ et al. (1991) Am J Med Genet 41:319–321]

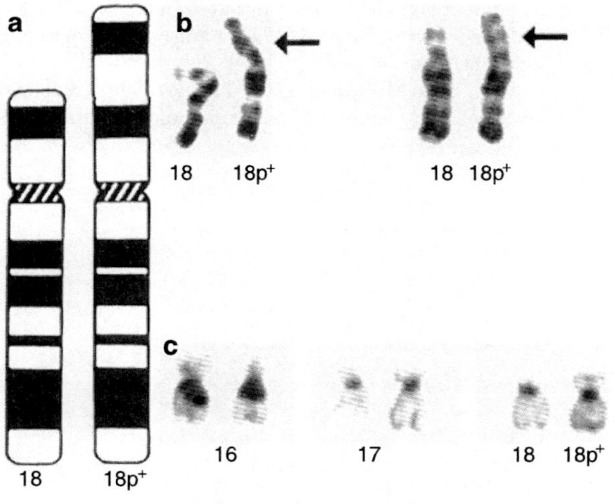

References

1. Craig-Holmes AP, Shaw MW (1973) Polymorphism of human C-band heterochromatin. Science 174:702–704
2. McKenzie WH, Lubs HA (1975) Human Q and C chromosomal variations: distribution and incidence. Cytogenet Cell Genet 14:97–115
3. Müller HJ, Klinger HP, Glasser M (1975) Chromosome polymorphism in a human newborn population. II. Potentials of polymorphic chromosome variants for characterizing the ideogram of an individual. Cytogenet Cell Genet 15(4):239–255
4. Lubs HA, Patil SR, Kimberling WJ, Brown J, Cohen M Gerald P, Hecht F, Myrianthopoulos N, Summit RL (1977) Q and C-banding polymorphisms in 7 and 8 year old children: racial differences and clinical significance. In: Hook EB, Porter IH (eds) Population cytogenetic studies in humans, Academic Press, New York, pp 133–159
5. Pittalis MC, Santarini L, Bovicelli L (1994) Prenatal diagnosis of a heterochromatic 18p+ heteromorphism (letter). Prenat Diagn 14:72–73
6. Zelante L, Notarangelo A, Dallapiccola B (1994) The 18ph+ heteromorphism. Prenat Diagn 14:1096–1097
7. Sensi A, Giunta C, Bonfatti A, Gruppioni R, Rubini M, Fontana F (1994) Heteromorphic variant 18ph+ analyzed by sequential CBG and fluorescence *in situ* hybridization. Hum Hered 44:295–297
8. Bonfatti A, Giunta C, Sensi A, Gruppioni R, Rubini M, Fontana F (1993) Heteromorphism of human chromosome 18 detected by fluorescent *in situ* hybridization. Eur J Histochem 37:149–154
9. Beverstock GC, Klumper F, Helderman V, Enden AT (1997) Yet another variation on the theme of chromosome 18 heteromorphisms? Prenat Diagn 17:585–586
10. Babu A, Verma RS (1986) The heteromorphic marker on chromosome 18 using restriction endonuclease AluI. Am J Hum Genet 38:549–554
11. Babu A, Verma RS, Patil SR (1987) AluI-resistant chromatin of chromosome 18: classification, frequencies and implications. Chromosoma 95:163–166
12. Wolff DJ, Raffel LJ, Ferre MM, Schwartz S (1991) Prenatal ascertainment of an inherited dup(18p) associated with an apparently normal phenotype. Am J Med Genet 41:319–321
13. Moog U, Engelen JJ, de Die-Smulders CE, Albrechts JC, Loneus WH, Haagen AA, Raven EJ, Hamers AJ (1994) Partial trisomy of the short arm of chromosome 18 due to inversion duplication and direct duplication. Clin Genet 46:423–429
14. Abeliovich D, Dagan J, Levy A, Steinberg A, Zlotogora J (1993) Isochromo-some 18p in a mother and her child. Am J Med Genet 46:392–393
15. Johnasson B, Mertens F, Palm L, Englesson I, Kristofferson U (1988) Duplication 18p with mild influence on phenotype. Am J Med Genet 29:871–874
16. Singer TS, Kohn G, Yatziv S (1990) Tetrasomy 18p in a child with trisomy 18 phenotype. Am J Med Genet 36:144–147
17. Pinto MR, Silva ML, Ribeiro MC, Pina R (1998) Prenatal diagnosis of mosaicism for tetrasomy 18p: Cytogenetic, FISH and morphological findings. Prenat Diagn 18:1095–1097

Chapter 24
Chromosome 19

Little attention has been paid to variations in C-banding of 19h [1]. McKenzie and Lubs [2] gave a frequency for 19cen+ as 1–2%. Other larger studies [3, 4] do not mention 19h. However, Crossen [5] noted four different classes of variants for 19h: (1) confined to the centromere; (2) extending into the short arm; (3) extending into the long arm; (4) extending into both long and short arms (examples, Fig. 24.1b, c). Verma and Luke [6] studied the incidence of the four classes using *Alu* digestion followed by Giemsa staining and gave incidences of 26, 17, 51 and 6%, respectively. Besides, variations in size and location of 19h, several reports

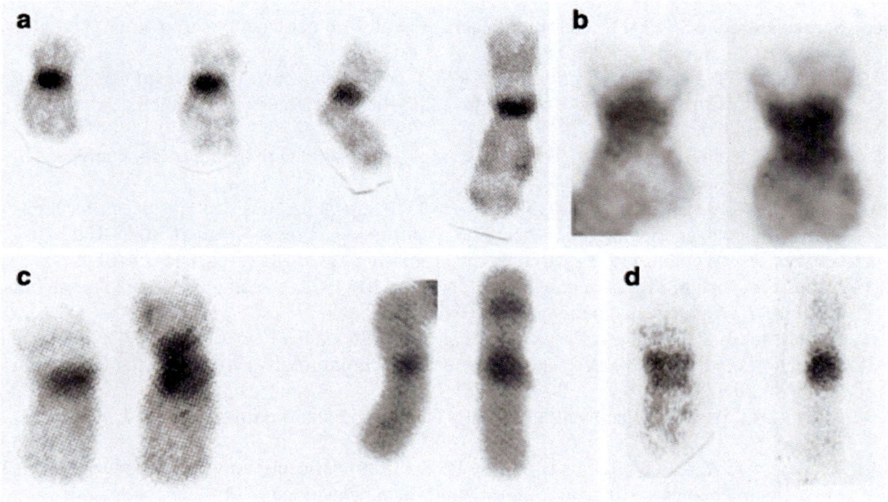

Fig. 24.1 **a** Normal chromosomes 19 from four different subjects by G-banding showing increasing band resolution. **b–d** Variant 19's from three different subjects showing partial inversions and duplications in centromeric heterochromatin by G- and C-banding, respectively. [Contributors: **a, b** Center for Human Genetics, Boston University (c1); **c, d** Jacqueline Schoumans, Haakland University Hospital (c24)]

H. E. Wyandt, V. S. Tonk, *Human Chromosome Variation: Heteromorphism and Polymorphism,* DOI 10.1007/978-94-007-0896-9_24,

[5–12] described "rare" pericentric inversions including inv(19)(p13q12), inv(19)(p11q13) and inv(19)(p13q13). Most were ascertained through an abnormal proband, but considered to be unrelated to the abnormal phenotype, and were not associated with increased abortion rate or lower reproductive fitness [11, 13, 14]. However, D'Alessandro et al. [12], describe three families with 19h inversions, inv(19)(p13q13) that all showed a "high frequency of abortions" as well as one unexplained perinatal death. A reported paracentric inversion, inv(19)(p13.2p13.3), ascertained through a proband with developmental delay and mild facial anomalies, was present in a normal mother and normal maternal grandfather [15].

References

1. Gardner HA, Wood M (1979) Variation in chromosome 19. J Med Genet 16:79–80
2. McKenzie WH, Lubs HA (1975) Human Q and C chromosomal variations: distribution and incidence. Cytogenet Cell Genet 14:97–115
3. Müller HJ, Klinger HP, Glasner M (1975) Chromosome polymorphism in a human newborn population II. Potentials of polymorphic chromosome variants for characterizing the idiogram of an individual. Cytogenet Cell Genet 15:239–235
4. Verma RS, Luke S, Brennan JP, Mathews T, Conte RA, Macera MJ (1993) Molecular topography of the secondary constriction region (qh) of the human chromosome 9 with an unusual euchromatic band. Am J Hum Genet 52:981–986
5. Crossen PE (1975) Variation in the centromeric banding of chromosome 19. Clin Genet 8:218–222
6. Verma RS, Luke S (1991) Heteromorphisms of pericentromeric heterochromatin of chromosome 19. Genet Anal Tech Appl 8:179–180
7. Sutherland GR, Gardiner AJ, Carter RF (1976) Familial pericentric inversion of chromosome 19, inv(19)(p13q13) with a note on genetic counseling of pericentric inversion carriers. Clin Genet 10:54–59
8. Singer TS, Kohn G, Yatziv S (1990) Tetrasomy 18p in a child with trisomy 18 phenotype. Am J Med Genet 36:144–147
9. Pinto MR, Silva ML, Ribeiro MC, Pina R (1998) Prenatal diagnosis of mosaicism for tetrasomy 18p: cytogenetic, FISH and morphological findings. Prenat Diagn 18:1095–1097
10. Gardner HA, Wood M (1979) Variation in chromosome 19. J Med Genet 16:79–80
11. Jacobs PA, Buckton KE, Cunningham C, Newton M (1974) An analysis of breakpoints in structural rearrangements in man. J Med Genet 11:50–64
12. D'Alessandro E, DeMatteis Vaccaraella C, Lo Re ML, Cappa F, D'Alfonso A, Della Penna MR, Del Porto G (1988) Pericentric inversion in chromosome 19 in three families. Hum Genet 80:203–204
13. Jordan DK, Taysi K, Blackwell NL (1980) Familial pericentric inversion 19. J Med Genet 17:222–225
14. Tharapel AT, Ward JC, Wiggins L, Wilroy RS Jr (1986) Pericentric inversion of chromosome 19: prenatal diagnosis and genetic counseling. Prenat Diagn 6:75–78
15. Phelan MC, Schroer RJ, Krug EF (1989) Paracentric inversion of chromosome 19 in three generations. Am J Med Genet 34:525–527

Chapter 25
Chromosome 20

Reports of variants of the centromeric region of chromosome 20 are rare. Only one case with 20ph+ is reported in a couple with recurrent miscarriages [1]. A second case was submitted to us as a normal variant (Fig. 25.1b), and a third case was detected in a patient with infertility [2]. In the latter case, the 20ph+ was positive by G- and C-banding and by FISH with alpha satellite probe D20Z1. In addition to

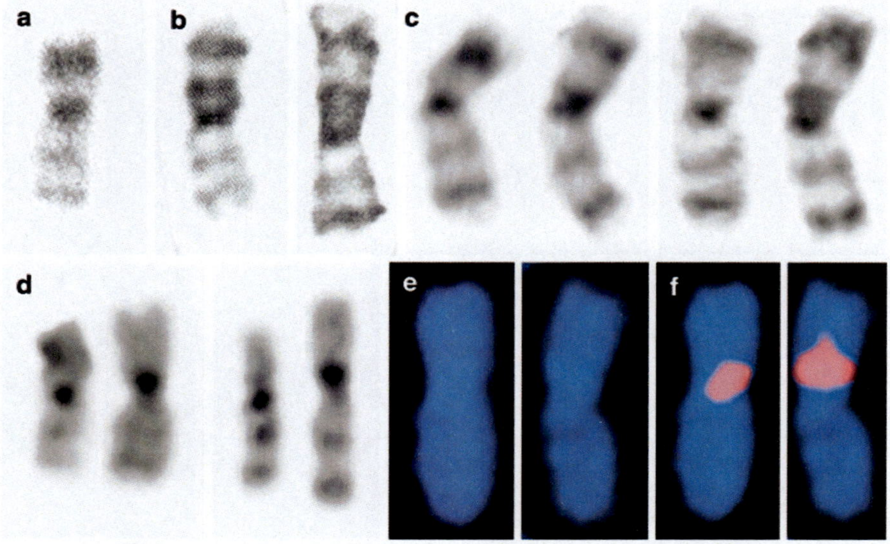

Fig. 25.1 **a** Normal chromosomes 20 from five different subjects by G-banding showing increasing band resolution. **b** Chromosomes 20 from two different cells from one individual showing 20ph+ by GTG-banding. Chromosome is maternal in origin. Pairs of chromosomes 20 by GTG-banding (**c**), by CBG-banding (**d**), by DAPI staining (**e**), and by FISH (**f**) with alpha satellite probe D20Z1 (CEP20, Vysis, Downers Grove, IL). The chromosome at right of each pair has 20ph+. [Contributors: **a, c–f** Center for Human Genetics, Boston University (c1); **b** K Yelevarthi and J Zunich, Indiana University, NW Medical Center (c14)]

H. E. Wyandt, V. S. Tonk, *Human Chromosome Variation: Heteromorphism and Polymorphism,* DOI 10.1007/978-94-007-0896-9_25,

20ph+, pericentric inversion in 20h also appears to be rare, there being one reported prenatal case known to us [3]. Further reports on variants of the centromeric region of chromosome 20 are needed to know their significance.

References

1. Romain DR, Whyte S, Callen DF, Eyre HJ (1991) A rare heteromorphism of chromosome 20 and reproductive loss. J Med Genet 28:477–478
2. Wyandt HE (2004) Introducton. In: Wyandt HE, Tonk VS (eds) Atlas of human chromosome heteromorphisms. Kluwer, Dordrecht, pp 3–10
3. Petersen MB (1990) Rare chromosome 20 variants encountered during prenatal diagnosis. Prenatal Diagn 6:363–367

Chapter 26
Chromosome 21

As already shown for acrocentric chromosomes 13–15, great variability in size and staining of centromere, short arm, stalk and satellites also apply to chromosome 21 (Fig. 26.1–26.3). Numerous studies have attempted to show a relationship between striking variants such as large satellites (Fig. 26.2), double satellites, large or double NOR's and an increased risk for non-disjunction, leading to trisomy 21 in Down syndrome. One case of an extremely large C-band positive short arm on a chromosome 21 was reported in the mother of a child with t(21q21q) translocation Down syndrome [1], but the variant was not attributed to be a cause of the child's translocation trisomy.

Similarly, many studies attempted to show a higher frequency of satellite association in parents of Down syndrome children. However most studies showed no significant difference in the frequency of satellite association in parents of trisomy 21 children and controls [2–5].

Occasionally, large bright satellites have been shown to be due to a translocation between an acrocentric short arm and the distal end of a Y chromosome. A few cases involve chromosome 21. These can have a distinctive morphology from typical satellites and can be identified by FISH. They can also be distinguished by special techniques such as adding distamycin to cultures [6] or staining with distamycin A and DAPI (See Chromosomes 15 and 22).

Q-band variants of chromosome 21 and other acrocentric chromosomes, have been useful in paternity studies and in determining parental origin of the extra chromosome 21 in trisomy 21 (Fig. 26.3) [7]. Occasionally, satellites that move around have been reported [8–10]. A case of an extreme variant that was unstable was reported by Livingston et al. [10] in a woman who had symptoms related to occupational exposure to organophosphate pesticides. Chromosome studies revealed a giant, Q-bright satellite that was present on a chromosome 21 in 15% of cells, on a chromosome 22 in 83% of cells, and on both 21 and 22 in 2% of cells. There was also a slightly increased frequency of sister chromatid exchange (SCE). The

H. E. Wyandt, V. S. Tonk, *Human Chromosome Variation: Heteromorphism and Polymorphism,* DOI 10.1007/978-94-007-0896-9_26,

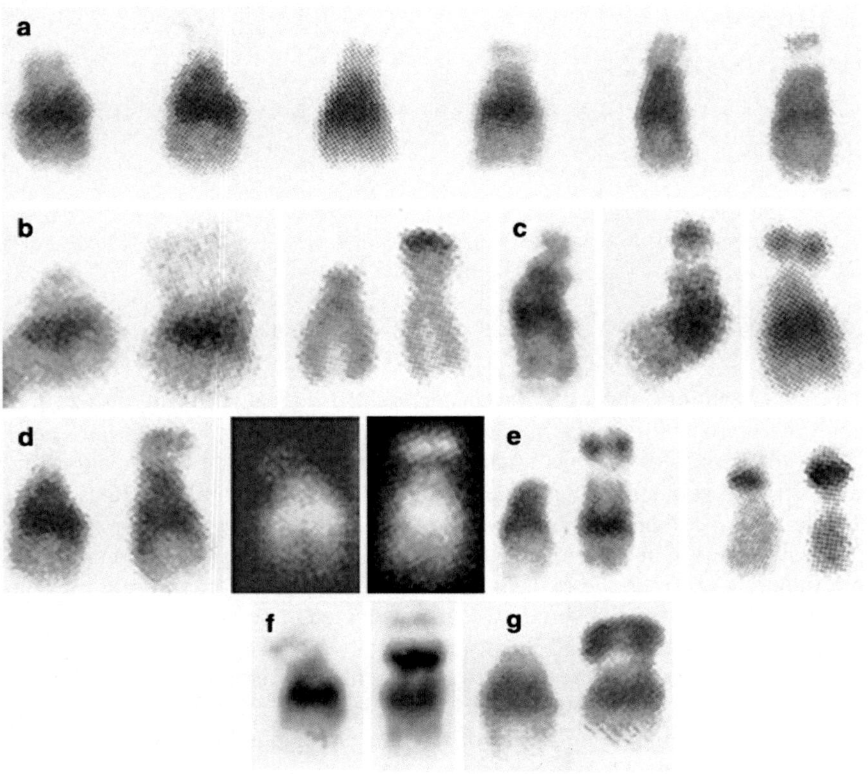

Fig. 26.1 **a** Chromosomes 21 from six normal subjects showing variations by GTG banding in the size and staining of the short arm without satellite (*far left*) and with satellite (*far right*). **b** Large (*right*) vs. typical (*left*) short arms of pair of chromosome 21's by GTG-banding. Pair of chromosomes by silver staining at right is from the same subject. **c** Chromosomes 21 by G-banding (*left*) from three normal subjects showing variations in size and intensity of staining of satellites. **d** Pair of chromosomes (*left*) by G-banding and from the same subject by Q-banding (*right*). **e** Pair of 21s stained by G-banding (*left*) showing large dark satellites and by silver staining (*right*) showing large NOR region (*far right*). **f** Pair of 21s by G-banding showing large dark short arm (*right*). **g** Pair of 21s by G-banding showing large dark satellite (*right*). [Contributor: Center for Human Genetics, Boston University (c1)]

woman's mother had the same giant satellite on a chromosome 22 in all of her cells. Three years post-exposure, a study of the affected woman's chromosomes revealed a lower SCE rate, but a similar distribution of the giant satellite to chromosomes 21 and 22.

Other notable variants of acrocentric chromosomes are those that show an apparent absence of heterochromatin or a short arm. Mayer et al. [1], in a study of C-band variants in 516 patients with mental retardation, found one case with 21p- and trisomy 21. In a previous study of 1,300 patients they found only one other

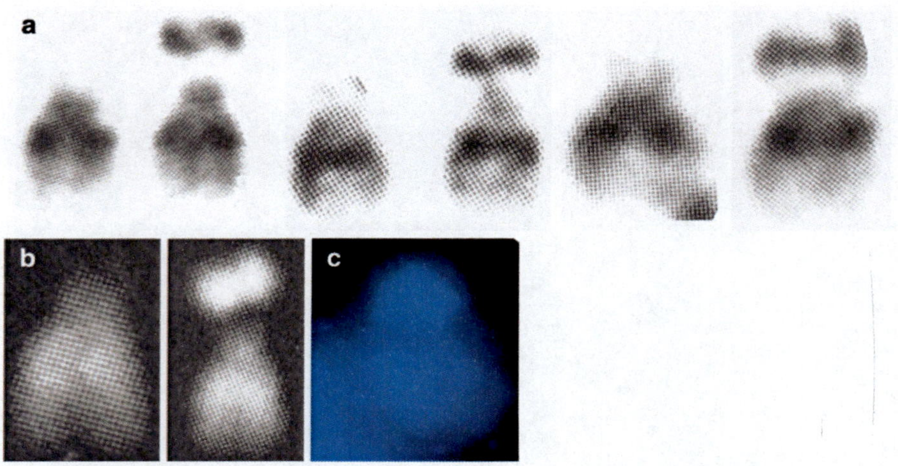

Fig. 26.2 **a** Three pairs of chromosomes 21 from the same subject showing large dark satellite by G-banding. **b** Pair of chromosomes 21 from the same subject showing intense staining by Q-banding. **c** Chromosome 21 from same subject showing large satellite to hybridize with alpha satellite probe, DYZ1 (Vysis, Downers Grove, IL), indicating its origin from the distal end of the Y long arm. [Contributor: Center for Human Genetics, Boston University (c1)]

case of 21p-, also in a family with Down syndrome due to t(21q21q). Instances of 21p- associated with Down syndrome have also been reported [11–14], but no causal relationship has been shown. It is possible that centromeric function could be affected in some of these cases if such a chromosome arose from a broken dicentric chromosome. The incidence of cases with 21p- or absent heterochromatin is also of interest in prenatal detection of trisomy in non-cultured amniocytes by FISH. Several reported cases of 21 aneuploidy have escaped detection by FISH, using alpha satellite probes, because of a small or absent signal [15–17]. According to Verma et al. [18] and Lo et al. [19] 3–3.7% of chromosome 21's do not show a hybridization signal, compared to 0.12% of chromosomes 13 and 17, and 0% of all other chromosomes.

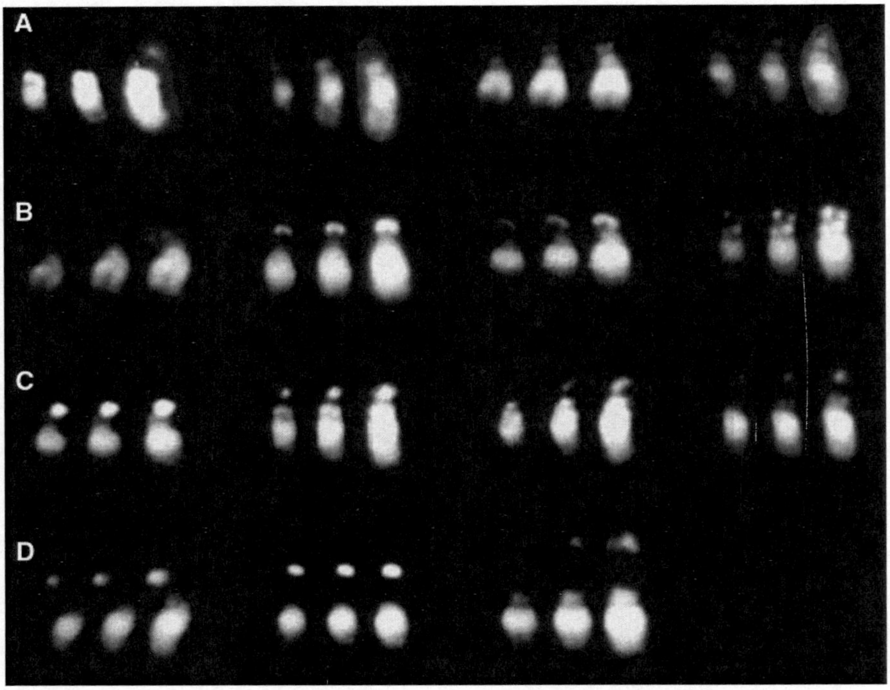

Fig. 26.3 Fifteen different variants of chromosome 21 from a population of 39 unrelated people. Each chromosome is serially printed to reveal heteromorphisms not visible at an exposure generally chosen to define the overall banding pattern. [Reproduced from Olson SB et al. (1986), Am J Hum Genet 38: 235–252]

References

1. Mayer M, Matsuura J, Jacobs, P (1978) Inversions and other unusual heteromorphisms detected by C-banding. Hum Genet 45:43–50
2. Cooke P, Curtis DJ (1974) General and specific patterns of acrocentric association in parents of mongol children. Humangenetik 23:279–287
3. Taysi K (1975) Satellite association: giemsa bandinng studies in parents of Down's syndrome patients. Clin Genet 8:319–323
4. Yip MY, Fox DP (1981) Variation in pattern and frequency of acrocentric association in normal and trisomy 21 individuals. Hum Genet 59:14–22
5. Jacobs PA, Mayer M (1981) The origin of human trisomy: a study of heteromorphisms and satellite associations. Ann Hum Genet 45:357–365
6. Spowart G (1979) Reassessment of presumed Y/22 and Y/15 translocations in man using a new technique. Cytogenet Cell Genet 23:90–94
7. Olson SB, Magenis RE (2004) Technical variables and the use of heteromorphisms in the study of human chromosomes. In: Wyandt HE, Tonk VS (eds) Atlas of human chromosome heteromorphisms. Kluwer, Dordrecht, pp 63–73
8. Gimelli G, Porro E, Santi F, Scappaticci S, Zuffardi O (1976) "Jumping" satellites in three generations: a warning for paternity tests and prenatal diagnosis. Hum Genet 34:315–318

9. Farrell SA, Winsor EJ, Markovik VD (1993) Moving satellites and unstable chromosome translocations: clinical and cytogenetic implications. Am J Med Genet 46:715–720
10. Livingston GK, Lockey JE, Witt KS, Rogers SW (1985) An unstable giant satellite associated with chromosomes 21 and 22 in the same individual. Am J Hum Genet 37:553–560
11. Tuncbilek E, Bobrow M, Clarke G, Taysi K (1976) A giant short arm of no. 21 chromosome in mother of 21/21 translocation mongol. J Med Genet 13:411–412
12. Shaw M (1962) Familial mongolism. Cytogenetics 1:141–179
13. De Grouche J (1970) 21p- maternal en double exemplaire chez un trisomique 21(French). Ann Genet (Paris) 13:52–55
14. Ballantyne GH, Parslow MI, Veale AM, Pullon DH (1977) Down's syndrome and the deletion of short arms of a G chromosome. J Med Genet 14:147–150
15. Lebo RV, Flandermeyer RR, Diukman R, Lynch ED, Lepercq JA, Golbus MS (1992) Prenatal diagnosis with repetitive *in situ* hybridization probes. Am J Med Genet 43:848–854
16. Bossuyt PJ, Van Tienen MN, De Gruyter L, Smets V, Dumon J, Wauters JG (1995) Incidence of low-fluorescence alpha satellite region on chromosome 21 escaping detection of aneuploidy at interphase by FISH. Cytogenet Cell Genet 68:203–206
17. Conte RA, Mathews T, Kleyman SM, Verma RS (1996) Molecular characterization of 21p-variant chromosome. Clin Genet 50:103–105
18. Verma RS, Batish SD, Gogineni SK, Kleyman SM, Statda DG (1997) Centromeric alphoid DNA heteromorphisms of chromosome 21 revealed by FISH-technique. Clin Genet 51:91–93
19. Lo AWI, Liao GCC, Rocchi M, Choo KHA (1999) Extreme reduction of chromosome-specific a-satellite array is unusually common in human chromosome 21. Genome Res 9:895–908

Chapter 27
Chromosome 22

Variations in chromosome 22 morphology and staining by different banding techniques (Fig. 27.1) are similar to other acrocentric chromosomes. Q-bright variants on chromosome 22 such as bright satellites are similar in frequency. However, bright variants of the short arm are more frequent. As with other acrocentric chromosomes variants, Q-band variants of 22 were useful in paternity studies (Fig. 27.2). Very large bright satellites are rare, but exchange with distal Yq occurs more frequently for chromosomes 15 and 22 than for the other acrocentric chromosomes. Special techniques such as staining with distamycin A and DAPI [1, 2] are useful to distinguish Q-bright Y chromatin from other Q-bright heterochromatin. Distamycin A, if added to cultures, causes under-condensation of heterochromatin originating from Yqh [3–5]. Currently, FISH using a probe specific for the satellite (DYZ1) sequences in Yqh is a quick, definitive test.

Translocations involving the repetitive sequences of Yqh region and acrocentric short arms are usually familial and of little or no clinical consequence. However, a rare dicentric translocation between a Y and an acrocentric reportedly resulted in malsegregation at meiosis and oligospermia [6]. Other rare occurrences that involve an unstable satellite that moved from chromosome 22 to other chromosomes were mentioned in the summary for chromosome 21 [7, 8], including one case of a prominent satellite moving from a t(10;22) to a normal 22 [9].

The pericentromeric region and short arm of chromosome 22 consist of several types of repetitive DNA including contiguous arrays of 1-, 3- and β-satellite [10–13]. The detailed molecular analysis of striking variants of heteromorphisms of the acrocentric chromosomes has not been done on a large scale. However, two cases of very large short arms on 22 were shown by Conte et al. [14] to have quite different organization of repetitive sequences. One had a tandem duplication of p11.2–p11.3. The second variant had lost β-satellite and ribosomal DNA regions and showed amplification of satellite III. At least one group of investigators has shown the presence of expressed sequences within the heterochromatin of chromosome 22 [15].

Chromosomes 14 and 22 share alpha satellite sequences, so that the centromeres of these two chromosomes cannot be distinguished by FISH. Several investigators have also shown cross hybridization of 13/21 alpha satellite DNA to a chromosome

H. E. Wyandt, V. S. Tonk, *Human Chromosome Variation: Heteromorphism and Polymorphism*, DOI 10.1007/978-94-007-0896-9_27,

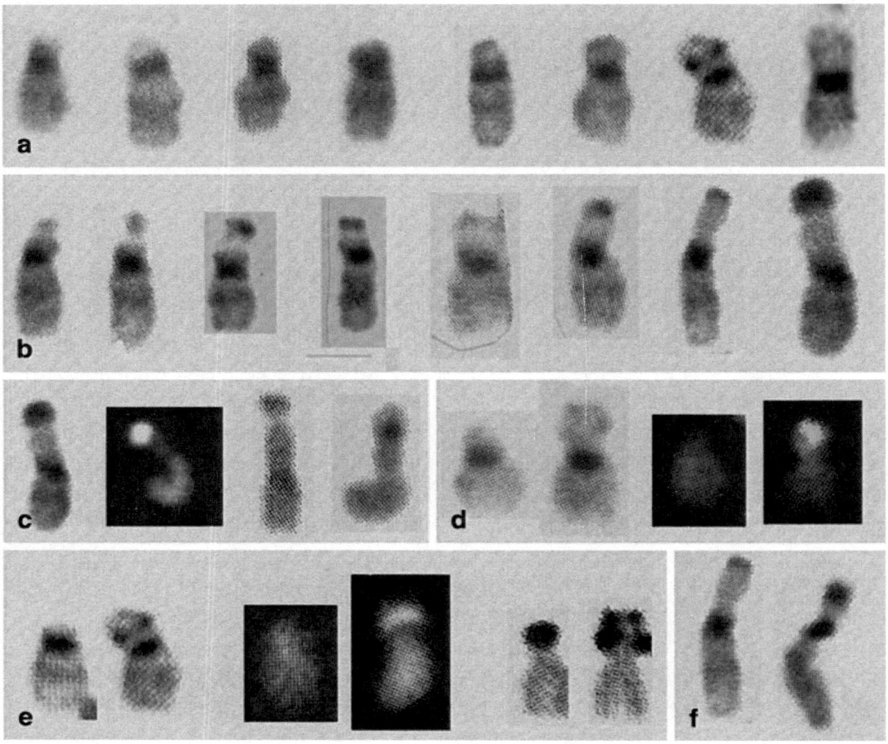

Fig. 27.1 **a** Chromosomes 22 by G-banding from eight different subjects showing variation in size of the short arm. **b** Chromosomes 22 from eight additional subjects showing variation in size and staining of the satellites and short arms. **c–f** Additional banding techniques on selected chromosomes: **c** chromosome with very long short arm, stalk and satellite by (*left to right*) G, Q-, C- and NOR-staining. Note lack of C-banding except at the centromere and on Q-bright satellite. A constriction just below the satellite corresponds to the NOR (*far right*). **d** Pairs of homologs from the same subject by G-banding (*left*) and by Q-banding (*right*). Note apparent lack of a *dark satellite* by G-banding and *very bright terminal satellite* by Q-banding (*right-hand chromosome of each pair*). **e** Pairs of homologs from the same subject by G-banding (*left*), Q-banding (*middle*) and NOR-staining (*right*). Note *dark band* in the middle of short arm is Q-bright and is flanked by two NOR regions. **f** Chromosome 22 with long pale-staining short arm by G-banding (*left*) and two NORs (*right*). [Contributor: **a–f** Center for Human Genetics, Boston University (c1)]

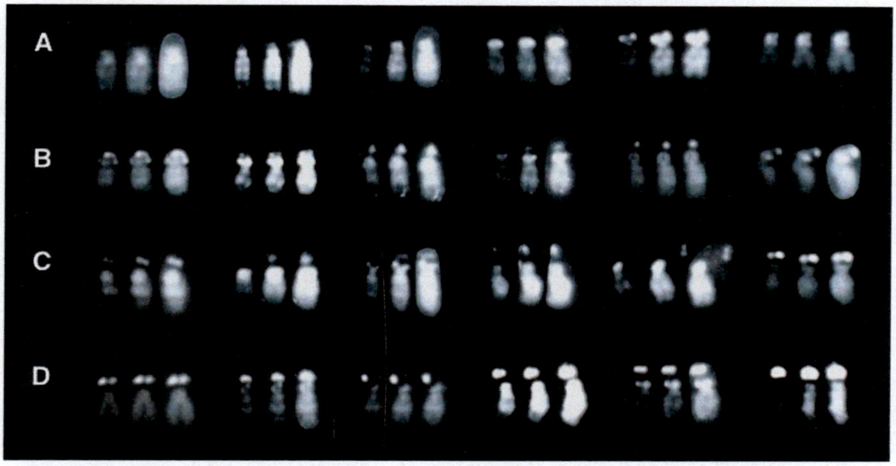

Fig. 27.2 Twenty four different variants of chromosome 22 from a population of 39 unrelated people. Each chromosome is serially printed to reveal heteromorphisms not visible at an exposure generally chosen to define the overall banding pattern. The *large bright satellites* in D-6 are relatively uncommon and were present in only one person of the population selected. [Reproduced from Olson SB et al. (1986), Am J Hum Genet 38:235–252]

22 by FISH, emphasizing the need for caution in using repetitive probes for screening for the common aneuploidies by interphase FISH [16–18].

References

1. Spowart G (1979) Reassessment of presumed Y/22 and Y/15 translocations in man using a new technique. Cytogenet Cell Genet 23:90–94
2. Funderburk SJ, Klisak I, Sparks RS, Carrel RE (1982) Familial Y-autosome translocation in two unrelated girls. Ann Genet 25:119–122
3. Schmid M (1979) Demonstration of Y/autosomal translocations using distamycin A. Hum Genet 53:107–109
4. Cohen MM, Frederick RW, Balkin NE, Simpson SJ (1981) Identification of Y chromosome translocations following distamycin A treatment. Clin Genet 19:335–342
5. Schmid M, Schmidtke J, Kruse K, Tolsdorf M (1983) Characterization of a Y/15 translocation by banding methods, distamycin A treatment of lymphocytes and DNA restriction endonuclease analysis. Clin Genet 24:234–239
6. Alves C, Carvalho F, Cremades N, Sousa M, Barros A (2002) Unique Y/13 translocation in a male with oligozoospermia: cytogenetic and molecular studies. Eur J Hum Genet 10:467–474
7. Gimelli G, Porro E, Santi F, Scappaticci S, Zuffardi O (1976) "Jumping" satellites in three generations: a warning for paternity tests and prenatal diagnosis. Hum Genet 34:315–318
8. Livingston GK, Lockey JE, Witt KS, Rogers SW (1985) An unstable giant satellite associated with chromosomes 21 and 22 in the same individual. Am J Hum Genet 37:553–560
9. Farrell SA, Winsor EJ, Markovik VD (1993) Moving satellites and unstable chromosome translocations: clinical and cytogenetic implications. Am J Med Genet 46:715–720

10. Rocchi M, Archidiacono N, Antonacci R, Finelli P, D'Aiuto L, Carbone R, Lindsay E, Baldini A (1994) Cloning and comparative mapping of recently evolved human chromosome 22-specific alpha satellite DNA. Somatic Cell and Mol Genet 20:443–448
11. Antonacci R, Rocchi M, Archidiacono N, Baldini A (1995) Ordered mapping of three alpha satellite DNA subsets on human chromosome 22. Chromosome Res 3:124–127
12. Mullenbach R, Pusch C, Holzmann K, Suijkerbuijk R, Blin N (1996) Distribution and linkage of repetitive clusters from the heterochromatic region of human chromosome 22. Chromosome Research 4:282–287
13. Shields C, Coutellle C, Huxley C (1997) Contiguous arrays of satellites 1, 3, and beta form a 1.5 Mb domain on chromosome 22p. Genomics 44:35–44
14. Conte RA, Kleyman SM, Laundon C, Verma RS (1997) Characterization of two extreme variants involving the short arm of chromosome 22: are they identical? Ann Genet 40:145–149
15. Blin N, Scholz M, Wissinger B, Mullenbach R, Pusch C (1997) Expressed sequences within pericentromeric heterochromatin of human chromosome 22. Mamm Geneome 8:859–862
16. Verlinsky Y, Ginsberg N, Chmura M, Freideine M, White M, Strom C, Kuliev A (1995) Cross-hybridization of the chromosome 13/21 alpha satellite DNA probe to chromosome 22 in the prenatal screening of common chromosomal aneuploidies by FISH. Prenat Diagn 15:831–834
17. Biancato JK (1996) Re: cross-hybridization of the chromosome 13/21 alpha satellite DNA probe to chromosome 22 in the prenatal screening of common aneuploidies by FISH (letter). Prenat Diagn 16:769–770
18. Tardy EP, Toth A (1997) Cross-hybridization of the chromosome alpha satellite DNA to chromosome 22 or a rare polymorphism? Prenat Diagn 17:487–488

Chapter 28
Chromosome X

McKenzie and Lubs [1] reported one Xcen+ variant was in a child whose parents were Mexican-American. In the New Haven study [2], 6 cases with Xcen+ were found: five were Caucasian; one was black. Friedrich et al. [3] did a blind determination of X centromere origin (by C-band size and location in the centromere, in Xp or in Xq) in 22 girls with Turner syndrome, compared with assignment of parental origin by RFLP analysis. They were able to trace the X centromere to the mother in 19 cases, in total agreement with RFLP analysis. Recently, a false-positive diagnosis of monosomy X by FISH was reported (Fig. 28.1) [4], due to a discrepancy in size of the alpha satellite signal, resulting in only one scorable signal in interphase.

Euchromatic variant: A female infant with Down syndrome (+21) and her normal mother had a non-mosaic interstitial deletion of band Xq26 [5]. The finding is of interest not only as a euchromatic variant in females, but it helps define a proximal limit of Xq deletion resulting in clinical features of Ullrich-Turner syndrome.

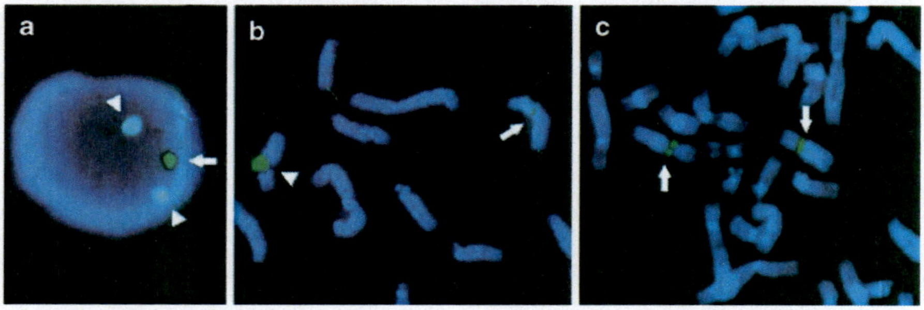

Fig. 28.1 Representative FISH images from the proband and mother. **a** Interphase FISH of uncultured amniocytes hybridized with chromosome 18 (*aqua*), X (*green*) and Y alpha satellite probes demonstrating two 18 (*aqua*) signals (*arrowheads*), but only one X signal (*arrow*). **b** Metaphase from cultured amniocytes hybridized with a BAC probe located within an X alpha satellite probe, showing a typical X chromosome signal (*arrowhead*), and a markedly reduced X signal (*arrow*). **c** Metaphase FISH from cultured amniocytes with a BAC probe located within a few hundred kilobases of DXZ1 on the long arm. [Reproduced with permission from Tsuchiya K et al. (2001), Prenat Diagn 21:852–855]

H. E. Wyandt, V. S. Tonk, *Human Chromosome Variation: Heteromorphism and Polymorphism,* DOI 10.1007/978-94-007-0896-9_28,

References

1. McKenzie WH, Lubs HA (1975) Human Q and C chromosomal variations: distribution and incidence. Cytogenet Cell Genet 14:97–115
2. Lubs HA, Patil SR, Kimberling WJ, Brown J, Cohen M, Gerald P, Hecht F, Myrianthopoulos N, Summit RL (1977) Q and C-banding polymorphisms in 7 and 8 year old children: racial differences and clinical significance. In: Hook EB, Porter IH (eds) Population cytogenetic studies in humans. Academic Press, New York, pp 133–159
3. Friedrich U, Larsen TB, Nielsen J (1991) Diagnostic reliability of the cytogenetic centromere heteromorphism in the human X chromosome. Clin Genet 40:465–466
4. Tsuchiya K, Schueler MG, Dev VG (2001) Familial X centromere variant resulting in a false-positive prenatal diagnosis of monosomy Xby interphase FISH. Prenat Diagn 21:852–855
5. Taysi K (1983) Del(X) (q26) in a phenotypically normal woman and her daughter who also has trisomy 21. Am J Med Genet 14:367–372

Chapter 29
Chromosome Y

It was recognized early that the Y chromosome varied greatly in length (Figs. 29.1, 29.2) in normal, fertile men [1–4]. Variations in length of the Q-bright fluorescent segment of the long arm were quickly shown to account for most of the size differences [5–7]. In large Ys, the distal Q-bright region is often composed of two fluorescent bands that are especially visible in prophase [8, 9]. In very small Ys, the Q-bright region may be greatly diminished in size or even absent [10, 11]. Other variant forms of the Y in normal males include pericentric inversions and satellited Ys.

In a large study of 11,148 consecutive newborns, 0.52% of boys had a large Y [12]. Nielsen and Friedrich [13] studied the length of the Y chromosome in a random sample of newborn boys compared with Y length in criminal males, and suggested a correlation between the size of the Y and risk of criminality[1]. Several studies [12, 14, 15] suggested an increased rate of spontaneous abortion associated with Yq+. Patil and Lubs [14] found a five-fold higher frequency of long Y in male infants whose mothers had three or more spontaneous abortions than in male infants whose mothers had less than three abortions. The proportion of pregnancies ending in prior abortion was two-fold higher in mothers of infants with a long Y. Similar findings were reported by Genest [15]. Lack of a Q-bright region, is more likely to be associated with minor anomalies and infertility [16, 17].

Variations in Length: Ghosh and Singh [18] studied the Y chromosome in two Indian populations and noted that 5% of Rajput males and 3% of Punjabi males had a long Y. Verma and colleagues [19–21] compared Y length to the length of F-group chromosomes in cohorts of 100 (Caucasian, American black and East Indian) males. For Caucasians, Y/F indices were classified into five groups: I, Y/F < 0.8 (0%); II, 0.81–0.94 (15%); III, 0.95–1.09 (66.7%); IV, 1.1–1.23 (13.3%); V, > 1.23 (5%). The distribution for American blacks was: I (0%), II (3.3%), III (56.7%), IV (30%) and V (10%). For East Indians the distribution was: I (0%), II (1.42%), III (15.7%), IV (58.7%) and V (24.3%). Hence, the frequency of longer Y's (Groups IV and V) was highest for East Indians and lowest for Caucasians. Indices were

[1] The significance of XYY and Y duplications in autism and other psyosociol behavior problems is still a controversial topic, outside the scope of this review.

H. E. Wyandt, V. S. Tonk, *Human Chromosome Variation: Heteromorphism and Polymorphism,* DOI 10.1007/978-94-007-0896-9_29,

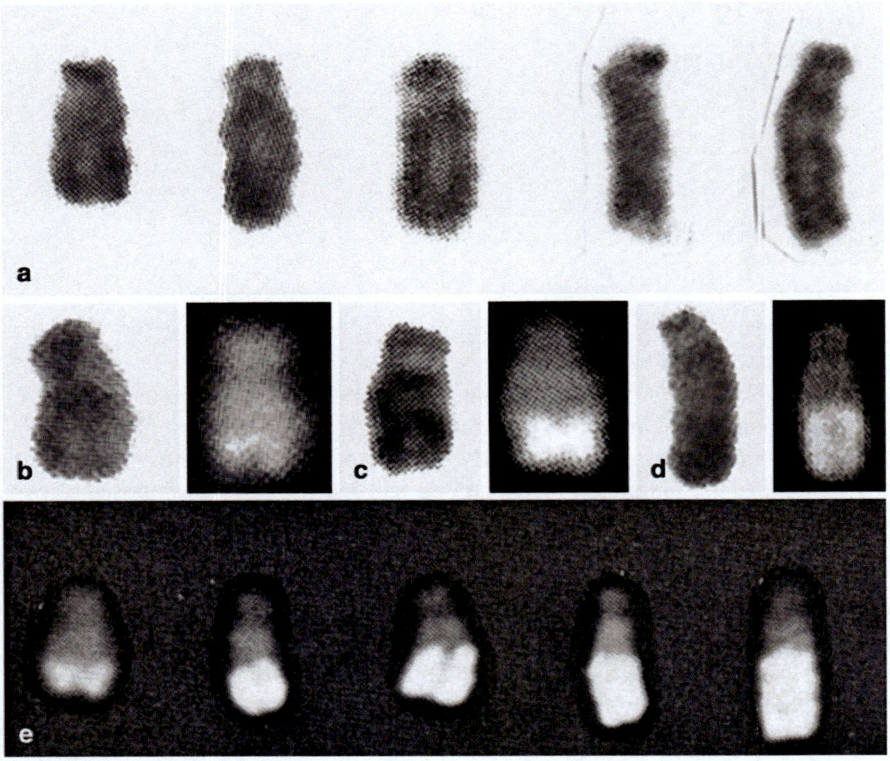

Fig. 29.1 **a** Y chromosomes from five different individuals showing increasing length of the qh region by G-banding. **b–d** Y chromosomes of different length from three different individuals showing comparison of qh regions by G-banding (*left*) and Q-banding (*right*). **e** Y chromosomes from a population of 39 unrelated individuals showing variations in fluorescent portion of the Y long arm with scores (for qh size) ranging from 1 (*left*) to 5 (*right*). [Part of Fig. 2, Olson et al. (1986), Am J Hum Genet 38:235–252]. [Contributors: **a–d** Center for Human Genetics, Boston University (c1)]

also calculated for the length of the non-fluorescent segment (nf)=nf/F. The non-fluorescent (nf/F) index was smallest for American blacks (range, 0.42–0.47) and largest for East Indians (range, 0.57–0.73), indicating longer Y's in East Indians also reflected an increase in size of the non-fluorescent (nf) segment, compared to black and Caucasian populations. Other studies of Asian populations have shown greater variation in the size of the Y than in the white population [22, 23], and have also shown that the non-fluorescent portion of the Y chromosome long arm varies in length [7, 24].

Inversion Y. The overall frequency of pericentric inversion, inv(Y) (p11q11), is approximately 1-2 per 1,000 [25], but varies in different populations. Only one case was found in a study of 14,835 consecutive newborns in a Japanese population [26]. On the other hand, inversion Y was found in 5.7% of a Gujarati Muslim Indian

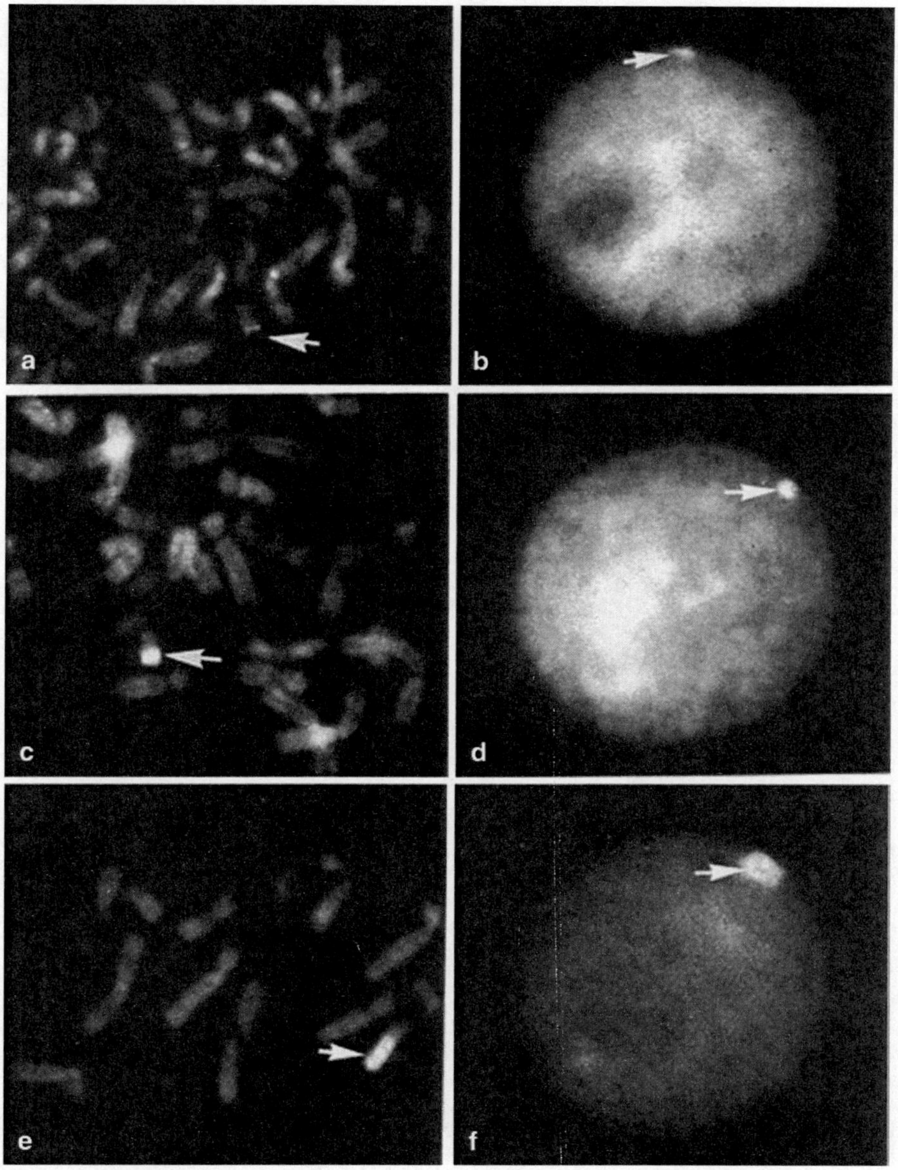

Fig. 29.2 Comparison of Yqh fluorescence of three different size Y chromosomes from three different individuals in metaphases (**a**, **c**, **e**) and in interphase cells (**b**, **d**, **f**). (H.E. Wyandt, unpublished)

population of South Africa [27], shown by molecular polymorphism studies [28] to have a common origin. In a study of a large Taiwanese population of 6,286 unrelated males, the frequency was 0.27%. Pericentric inversions in the Y (Fig. 29.3a–c) are generally considered to be normal variants with no apparent effect on fertility [19]. However, a handful of cases have been reported in infertility studies [29–31]. Two cases have been associated with asthenonecrozoospermia and oligospermia, respectively. Although the inversions appeared to be the standard type, DNA analysis revealed interstitial microdeletions [29]. One case with oligospermia and inv(Y)(p11q11) was revealed by FISH to have a break in Yq11, through the DAZ gene [31]. A few reports of multiple anomalies associated with aneuploidies [32, 33] or increased fetal wastage [34] appear to be spurious and probably coincidental.

Two cases of atypical Y rearrangements are noteworthy. Rivera et al. [35] report a three generation family with an unstable familial Y chromosome with a single Cd-positive constriction that occasionally assumed an acrocentric appearance, presumably due to an inversion. FISH with DYZ3 revealed a signal outside the primary constriction. Another case (Fig. 29.3a–e), interpreted to be a possible insertion, was detected prenatally and was also present in the father [Wyandt and Huang, unpublished]. The father had mosaicism, with a few cells showing Y loss and or two Ys. The two cases appear to be similar in that both showed alpha satellite signals away from the primary constriction, indicating that either alpha satellite is not necessary for centromere function or a new class of latent heteromorphism called a "neocentromere" was induced by the rearrangements. Neocentromeres are increasing being discovered at specific sites throughout the human genome [36].

Satellited Y: Familial satellited Y (Yqs) chromosomes (Fig. 29.4) were first reported by Genest et al. [37] and by Schmid [38]. Both were ascertained through abnormal probands, one with trisomy 21[39]. In the second family [40], two sisters with congenital heart defects and other anomalies died, and the father was shown to have both Yqs and a fragile17p. Genest et al. then reported a third family [39–41], also ascertained through a child with Down syndrome, in which the Yqs could be

Fig. 29.3 **a–c** Y chromosomes with pericentric inversions from three different individuals with pair at far right **c** showing the same Y chromosome by G and Q-banding. **d** Intrachromosomal insertion, ins(Y)(p11.2q10q11.23) by G-banding (*left*) and by Q-banding (2nd from *left*). FISH analysis with DYZ1/DYZ3 (Vysis) is shown (*middle*), with D15Z3 alone (2nd from *right*) and SRY (far *right*). (Xin Li Huang and H.E. Wyandt, unpublished). **e** Diagram of mechanism of formation of **d** with rearranged chromosome at far right showing results of FISH and banding analysis. (Xin Li Huang and H.E. Wyandt, unpublished). Chromosome, present in a fetus and in the father, was unstable in the father (see Y summary). Intact SRY region was also confirmed by molecular analysis. **f** Satellited, non-fluorescent Y chromosome studied by different FISH analyses using D15Z1, green (a, b), DYZ1, red and DYZ3, green (c), Y WCP, spectrum orange (d) and Xq/Yq telomeric associated sequence, red (e) probes on (**a**) the proband and (**b**) the father. The chromosome 15 specific classical satellite DNA probe (D15Z1) indicates chromosome 15 as the origin of the satellites.In **a** and **b** chromosomes 15 and Y of the proband and father come from the same metaphase. The negative signal for the Yq telomeric probe **c** suggests loss of Yqter on the Y. [Reproduced with permission from Verma et al. (1997), Characterization of a satellited non-fluorescent Y chromosome (Y[nfqs]) by FISH. J Med Genet 34:817–818. Contributor: **a–e** Center for Human Genetics, Boston University (c1)] traced back, patrilineарly, 10 generations (over three hundred years) [41]. Several additional cases were subsequently reported [42–44] and fourteen families were reviewed by Schmid et al. [38]. Yqs was frequently observed in satellite association and silver staining confirmed the presence NOR regions in all but one. Chromosomal variants revealed different breakpoints and independent origins for all fourteen families. All of the cases except one [44] were familial and transmitted patrilineарly through several generations. Despite frequent ascertainment through congenital ab-

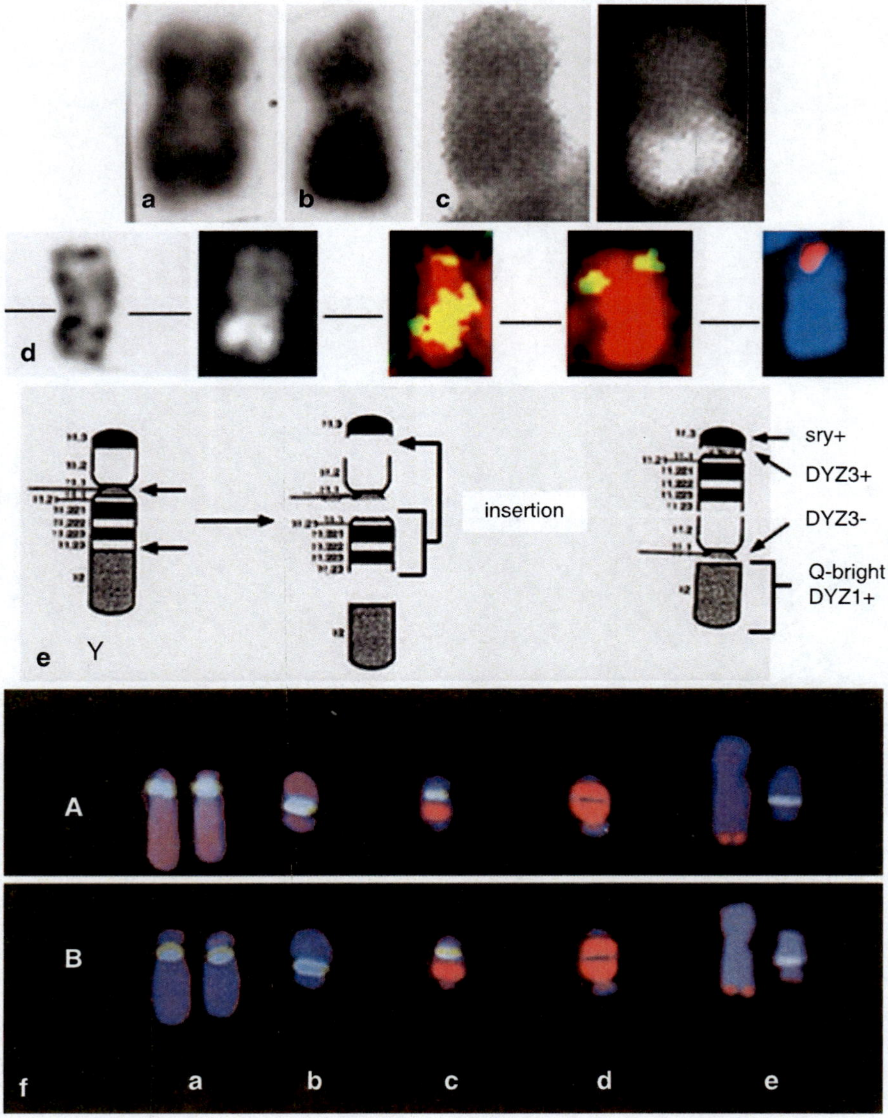

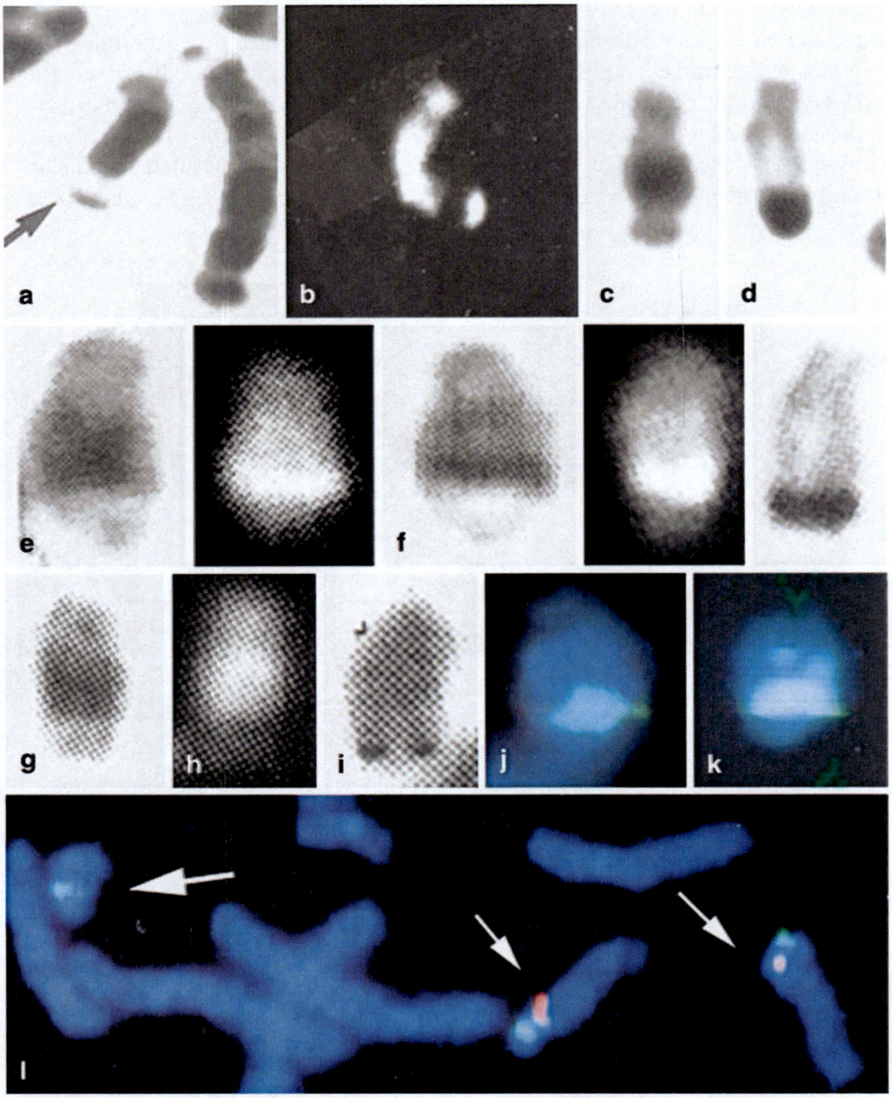

Fig. 29.4 **a–d** Familial satellited Y chromosome (B. Haddad and H.E. Wyandt, unpublished) by G-banding (*arrow*, **a**), Q-banding (**b**), C-banding (**c**) and silver staining (**d**). **e, f** Satellited Y from a different family (father and son) by G and Q banding (**e**, **f**) and silver staining also shown (**f**, far *right*). **g–l** Satellited Y in a third family (father and son) by G-banding (**g**), Q-banding (**h**), silver staining (**i**) and by FISH (**j**, **k**). Probes include alpha satellite DYZ1 for Yqh alone (**j**), DYZ1 and DYZ3 (alpha satellite specific for the centromeric region), and D15Z1 (satellite III specific to the short arm of chromosome 15) with probe for SNRPN in 15q11.2 (**l**) Probe for D15Z1 (*green*) present on the satellited Y chromosome is indicated by the large arrow. The small arrows indicate two chromosomes 15. Orange signals represent SNRPN. [Contributor(s): **e–k** Center for Human Genetics, Boston University (c1)]

normalities or other chromosome abnormality, most progeny of Yqs carriers were phenotypically normal. Verma et al. [45] describe a familial non-fluorescent satellited Y chromosome in a prenatal sample (Fig. 29.3f), with satellited material revealed by FISH to be from chromosome 15.

References

1. Cohen MM, Shaw MW, MacCluer JW (1966) Racial differences in the length of the human Y chromosome. Cytogenetics 5(1):34–52
2. El-Alfi O (1970) A family with a large Y chromosome. J Med Genet 7:37–40
3. Makino S, Muramoto T (1964) Some observations on the variability of the human Y chromosome. Proc Jap Acad 40:757–761
4. Bishop A, Blank CE, Hunter H (1962) Heritable variation in the length of the Y chromosome. Cytogenetics 5:34–52
5. Bobrow M, Pearson PL, Pike MC, El-Alfi OS (1971) Length variations in the quinacrine-binding segment of human Y chromosomes of different sizes. Cytogenetics 10:190–198
6. Laberge C, Gagne R (1971) Quinacrine mustard staining solve the length variations of the human Y chromosome. John Hopkins Med J 128:79–83
7. Schnedl W (1971) Fluoreszenzuntersuchugen uber die angenvariabiliatat des Y-Chromosoms beim Menschen. Humangenetik 12:188–194
8. Sperling K, Lackman I (1971) Large human Y chromosome with two fluorescent bands. Clin Genet 2:352–355
9. Kim MA, Bier L, Pawlowitzki IH, Pfeiffer RA (1971) Human Y chromosome with two fluorescing bands after staining with quinacrine derivates. Humangenetik 13:238–240
10. Borgoankar DS (1971) Non-fluorescent Y chromosome. Lancet 15:1017
11. Wahlstrom J (1971) Are variations in length of Y chromosome due to structural changes. Hereditas 69:125–128
12. Nielsen J (1978) Large Y chromosome (Yq+) and increased risk of abortion. Clin Genet 13:415–416
13. Nielsen J, Friedrich U (1972) Length of the Y chromosome in criminal males. Clin Genet 3:281–285
14. Patil SR, Lubs HA (1977) A possible association of long Y chromosomes and fetal loss. Hum Genet 35:233–235
15. Genest P (1979) Chromosome variants and abnormalities in 51 married couples with repeated spontaneous abortions. Clin Genet 16(6):387–389
16. Tishler PV, Lamborot-Manzur M, Atkins L (1972) Polymorphism of the human Y chromosome: fluorescence microscopic studies on the sites of morphologic variation. Clin Genet 3:116–122
17. Robinson JA, Buckton KE (1971) Quinacrine fluorescence of variant and abnormal human Y chromosomes. Chromosoma (Berl) 35:342–352
18. Ghosh PK, Singh IP (1975) Morphological variability of the human chromosomes in two Indian populations—Rajputs and Punjabs. Humangenetik 29(1):67–78
19. Verma RS, Evans-McCalla M, Dosik H (1982) Human chromosomal heteromorphism in American Blacks. VI. Higher incidence of longer Y owing to non-fluorescent (nf) segment. J Med Genet 19:297–301
20. Verma RS, Huq A, Dosik H (1983) Racial variation of a non-fluorescent segment of the Y chromosome in East Indians. J Med Genet 20:102–106
21. Verma RS, Dosik H, Scharf T, Lubs HA (1978) Length heteromorphisms of fluorescent (f) and non-fluorescent (nf) segments of human Y chromosome: classification, frequencies and incidence in normal Caucasians. J Med Genet 15:277–281

22. Hsu LYF, Benn PA, Tannenbaum HL, Perlis TE, Carlson AD (1987) Chromosome polymorphism of 1, 9 16 and Y in 4 major ethnic groups: a large prenatal study. Am J Med Genet 26:95–101
23. Hou JW, Wang TR (1999) Study of human Y chromosome polymorphism in Taiwan. Acta Paediatr Twaiwan 40:302–304
24. Soudek D, Laraya P (1974) Longer Y chromosome in criminals. Clin Genet 6:225–229
25. Shapiro LR, Pettersen RO, Wilmot PL, Warburton D, Benn PA, Hsu LYF (1984) Pericentric inversion of the Y chromosome and prenatal diagnosis. Prenat Diagn 4:463–465
26. Maeda T, Ohno M, Matsunobu A, Yoshihara K, Yabe N (1991) Cytogenetic survey of 14,835 consecutive liveborns. Jinrui Idengaku Zasshi 36:117–129
27. Bernstein R, Wadee A, Rosendorff J, Wessels A, Jenkins T (1986) Inverted Y chromosome polymorphism in the Gujarati Muslim Indian population of South Africa. Hum Genet 74:223–229
28. Spurdle A, Jenkins T (1992) The inverted Y chromosome polymorphism in the Gujarati Muslim Indian population of South Africa has a single origin. Hum Hered 42:330–332
29. Tomomasa H, Adachi Y, Iwabuchi M, Oshio S, Umeda T, Lino Y, Takano T, Nakahori Y (2000) Pericentric inversion of the Y chromosome of an infertile male. Arch Androl 45:181–185
30. Giltay JC, van Golde RJ, Kastrop PM (2000) Analysis of spermatozoa from seven ICSI males with constitutional sex chromosomal abnormalities by fluorescence *in situ* hybridization. J Assist Reprod Genet 17:151–155
31. Causio F, Canale D, Schonauer LM, Fischetto R, Leonetti T, Archidiacono N (2000) Breakpoint of a Y chromosome pericentric inversion in the DAZ gene area. A case report. J Reprod Med 45:591–594
32. Acar H, Cora T, Erkul I (1999) Coexistence of inverted Y, chromosome 15p+ and abnormal phenotype. Genet Couns 10:163–170
33. Motos GMA (1989) Pericentric inversion of the human Y chromosome. An Esp Pediatr 31:583–587
34. Toth A, Gaal M, Laszlo J (1984) Familial pericentric inversion of the Y chromosome. Ann Genet 27:60–61
35. Rivera H, Vassquez AI, Ayala-Madrigal ML, Ramirez-Duenas ML, Davalos IP (1996) Alphoidless centromere of a familial unstable inverted Y chromosome. Ann Genet 39:236–239
36. Warburton PE (2004) Chromosomal dynamics of human neocentromere formation. Chromosome Res 12:617–626
37. Genest P, Bouchard M, Bouchard J (1967) A satellited human Y chromosome: an evidence of autosome gonosome translocation. Canad J Genet Cytol 9:589–595
38. Schmid M, Haaf T, Solleder E, Schempp W, Leipoldt M, Heilbronner H (1984) Satellited Y chromosomes: structure, origin, and clinical significance. Hum Genet 67:72–85
39. Genest P (1978) A satellited Y chromosome. [In French]. Ann Genet 21:237–238
40. Genest P (1979) Remarks on a satellited Y chromosome. [In French]. Sem Hop 55:799–80
41. Genest P, Laberge C, Poty J, Gagne R, Bouchard M (1970) Transmission d'un petit "Y" durant onze generation dans une lignee familiale [In French]. Ann Genet (Paris) 13:233–238
42. Howard-Peebles PN, Stoddard GR (1976) A satellited Yq chromosome associated with trisomy 21 and an inversion of chromosome 9. Hum Genet 34:223–225
43. Bayless-Underwood L, Cho S, Ward B, Robinson A (1983) Two cases of prenatal diagnosis of a satellited Yq chromosome. Clin Genet 24:359–364
44. Turleau C, Chavin-Colin F, Seger J, Sorin M, Salet D, de Grouche J (1978) Satellited Y chromosome (Yqs) and nucleolar organizer occuring de novo. Ann Genet 21:239–242
45. Verma RS, Gogineni SK, Kleyman SM, Conte RA (1997) Characterization of a satellited non-fluorescent Y chromosome (Y[nfqs]) by FISH. J Med Genet 34:817–818

Chapter 30
FISH Variants

A handful of variants have been characterized by FISH analysis. The main impediments are (1) the expense of developing and characterizing probes that are not necessarily disease-related and (2) the application to sufficiently large populations to establish variant frequencies. While FISH is a valuable adjunct to standard cytogenetic techniques, care must be taken in interpreting results from probes that have not been adequately studied or characterized. New variants are emerging from techniques that purport to detect subtle duplications and deletions that may not be visible cytogenetically. These techniques include results from subtelomeric painting probes, from comparative genomic hybridization (CGH), either metaphase or array CGH, and likely other "locus-specific" probes that have not been fully characterized.

One example of the latter is a 28-kb deletion spanning the D15S63 locus first described by Buiting et al. [1] and confirmed by Silverstein et al. [2]. The frequency of the 28-kb deletion was 1/75 in Ashkenazi Jews, but is rarer in people of mixed origin. Since methylation analysis is often used with probe PW71 (D15S63) for Prader—Willi and Angelman syndromes, such a variant can lead to false positive results. The authors recommended the use of SNRPN rather than D15S63 in the diagnosis of PWS.

The number and types of variants revealed by FISH depends on the known specificity of the probe and whether a particular probe has been applied in the study of a significant number of normal individuals. Unfortunately, most commercially available probes are produced in small lots with a limited number of tests per lot possible by any one laboratory. Any suspected variant, therefore, is best confirmed by testing multiple individuals with overlapping lots of probe.

H. E. Wyandt, V. S. Tonk, *Human Chromosome Variation: Heteromorphism and Polymorphism,* DOI 10.1007/978-94-007-0896-9_30,

30.1 FISH Results with Centromeric Repeats

Certain commercially available centromeric probes hybridize to more than one chromosome (e.g. D13/21Z1 and D14/22Z1), hence limiting their utility in identifying the centromeres of these chromosomes. These probes are no longer sold by Vysis-Abbott. Such cross-hybridization should not be considered to be normal variants, because probed sequences are typically shared by these chromosomes in the majority of the population. On the other hand, D15Z1, a satellite III sequence, is usually specific for a variable number of sequences on the short arm of chromosome 15. As already discussed, a signal with D15Z1 is frequently seen on chromosome 14. Atypical results, such as this, or cases in which chromosome-specific sequences are absent from one chromosome or the other and cases in which the number of repeats deviates from the normal number to produce a larger or smaller signal than average are appropriate to be regarded as variants. The following are descriptions of cross-hybridization that occurs with some commercially available probes, due either to non-specificity of the probe or to a high frequency of atypical signals:

Chromosomes 1, 5 and 19: The commercial centromeric probe from Oncor (no longer available) cross-hybridized with chromosomes 1, 5 and 19. The commercial probe for chromosome 1 from Vysis, while reportedly specific for chromosome 1, occasionally also showed cross hybridization with 5 and/or 19. It was not always clear whether such cross-hybridization was a polymorphism or represented a contamination of probe by the closely related sequence. Satellite sequence D1Z5 is specific to the ph region of chromosome 1. Satellite sequences D1Z7, D5Z2 and D19Z3 represent an alphoid array that is shared by ph regions of 1, 5 and 19. A third alphoid array consisting of D5Z1 and D19Z2 is shared by the ph regions of 5 and 19 [3]. Probes which contain all three arrays will hybridize to all three chromosomes.

Chromosomes 13 and 21: Alpha satellite probes were initially used for rapid detection of aneuploidy in uncultured amniotic fluid cells. Verlinsky et al. [4] reported three false negative results and one false positive in 516 prenatal cases, using alpha satellite probe for chromosomes 13 and 21. False negatives were interpreted to be due either to failure of hybridization or to polymorphism. A diminished signal size in the chromosome 13 centromeric region of occasional individuals had also been previously reported by Weier and Gray [5] and Lapdot-Lifson et al. [6], who recommended caution in interpreting interphase FISH results to diagnose aneuploidy. Bossuyt et al. [7] gave a similar warning in a report of diminished signal intensity in four of 101 cases. The false-positive case of Verlinsky et al. [4] was due to cross-hybridization of the centromeric probe for 13 and 21 with a chromosome 22. Similar cross-hybridization with chromosome 22 also occurred in the mother. Tardy and Toth [8] reported similar cross-hybridization to the chromosome 22 of a 4-year-old boy with mild mental retardation and dysmorphism and of his unaffected father. Hybridization with unique probes for 14 and 22 failed to reveal a translocation between 13 or 21 and 14 or 22. Therefore, the investigators interpreted this cross-hybridization to represent a normal polymorphism. Such polymorphism is also supported by molecular studies. Vissel and Choo [9] reported four distinct alpha satellite subfamilies shared by chromosomes 13, 14 and 21.

Chromosomes 14 and 15: Two distinct classes of satellite DNA are found on chromosome 15 [10], alpha satellite and classical satellite. The classical satellite sequence, corresponding to satellite III and DA/DAPI positive regions on the short arm of chromosome 15 reportedly cross-hybridizes with short arm regions of other acrocentrics with a frequency of about 10%. Stergianou et al. [11] found cross-hybridization in 12 of 100 randomly selected individuals studied. This was consistent with a study by Smeets et al. [12] using non-fluorescent immunoperoxidase detection of alphoid sequences specific for chromosome 15. They found hybridization corresponding to DA/DAPI-positive regions on acrocentric chromosomes other than 15 in 7 of 127 individuals studied. In a FISH analysis of a case with 15p- (Fig. 30.1a), classical satellite D15Z1 (Vysis) was absent. However, alpha satellite D15Z4 (Vysis), specific for the centromeric region, was present [13]. In another case being tested for Prader-Willi syndrome from this same study (Fig. 30.1b), both chromosome 14 homologs showed a signal with D15Z1. A recent study by Cockwell et al. [14] of 1657 individuals with normal chromosomes and with acrocentric involvement revealed 1 in 6 individuals with additional signals of D15Z1 irrespective of whether a chromosome abnormality was present. Furthermore, D15Z1 was present in an additional copy, not only on chromosome 14, but on 13, 21 and 22 in decreasing size and frequency. Cases homozygous for a D15Z1 signal on chromosome 14 were also seen at approximately the expected frequency. However, the finding was considered uncommon enough that consideration of possible uniparental disomy for chromosome 14 was recommended in such cases. Two cases of chromosome 15 lacking the D15Z1 signal were also reported.

Earl et al. [15] found absence of satellite III DNA in the centromere and proximal long arm of a 14p- variant, but retention of the centromeric region. These results demonstrated that satellite III was not an essential component of centromere function but this does not exclude the possibility satellite DNA enhances and/or protects centromeric function.

Chromosome 18 and Other Rare Variants: Bonfatti et al. [16] found individual variations in pericentromeric regions of chromosome 18 with the alpha satellite probe that were almost undetectable in size by C-banding. In addition to the above cases, a number of alpha satellite variants are included in the present volume including variants of chromosomes 11, 13, 14, 15, 18 and 20 (see individual chromosomes).

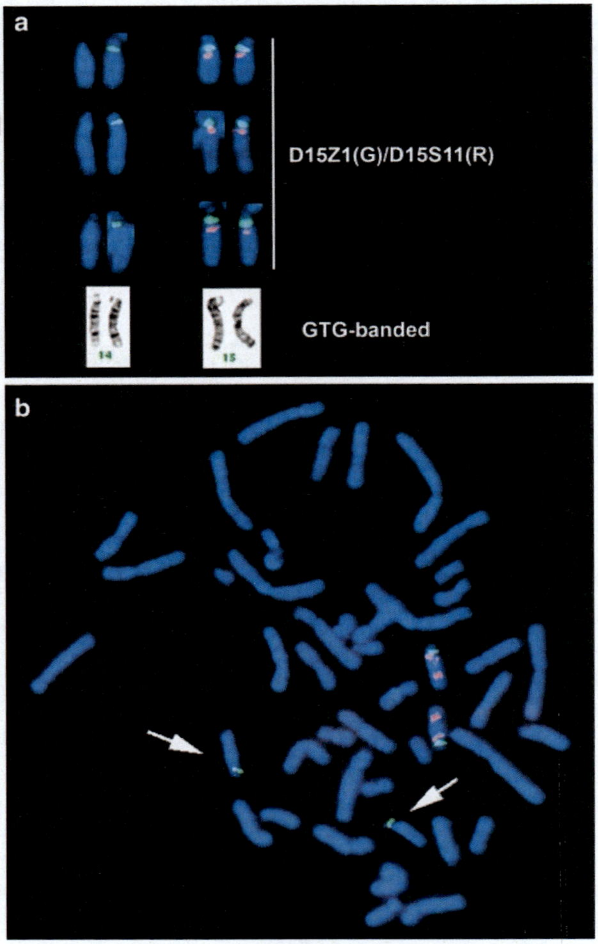

Fig. 30.1 **a** Apparently common variant of chromosome 14 seen with the chromosome 15-specific probe for satellite III sequences (D15Z1). Partial karyotypes of 14s and 15s from three different cases showing extra signal (*green*) with D15Z1 on the *short arm* of a chromosome 14 (*left hand* pair of chromosomes in each row). Chromsomes 15 (*right hand* pair in each row) shows *green* signal with D15Z1 and *orange* signal with probe for SNRPN in 15q11.2. Last row shows two normal appearing 14s by G-banding from one of the cases. **b** Metaphase showing two chromosomes 14 with a signal with D15Z1 (*arrows*). Two normal 15s have signals for D15Z1 (*green*), for SNRPN (*orange*) in proximal 15q and for PML (*orange*) in distal 15q

30.2 Subtelomeric Deletions/Duplications: Normal Variation or Chromosome Abnormality

Subtelomeric partial painting probes have been used to detect cryptic or semi-cryptic chromosome abnormalities in individuals with idiopathic dysmorphic features and/or mental retardation [17]. Depending on the population being tested, the frequency of abnormalities varies from 0.5% in cases with mild retardation to 7.4% in cases with severe retardation [18]. An additional factor in affecting these frequencies is whether or not the abnormality is truly cryptic. When a minimum resolution of 550 bands is imposed for karyotype detection, the frequencies of truly cryptic abnormalities are reduced by about half (Yu et al. [19]).

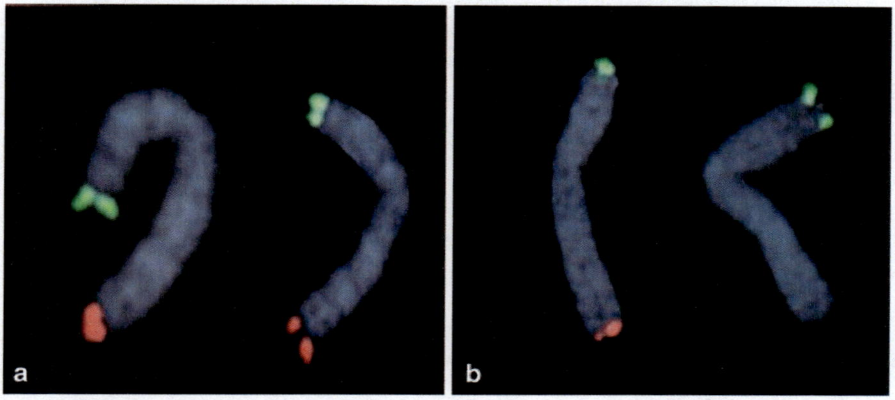

Fig. 30.2 Deletion variant at 2q telomeric region involving the locus D2S2986 was reported in a mentally retarded child and a normal father. The Vysis subtelomere probe **a** that was normal did not include D2S2986, but the Cytocell probe set that did include D2S2986 **b** was deleted in the child and father. The 2pter is *green* and 2qter is *orange*. This deletion variant occurs in about 8% of the population. [Repropduced from Plate 67 Atlas of Human Chromososme Heteromorphisms (Wyandt HE and Tonk V, eds), Kluwer, Dordrect, 2004, contributed by Jalal SM: reported by Jalal et al. (2000). Am J Hum Genet 67: A770 (c42)]

Probes used in subtelomeric studies have been tested through at least two generations of development. The first generation showed a high frequency of cross-hybridization [20] with other chromosome regions, indicating that varying degrees of sequence homology existed between different chromosomes. A second generation of probes showed less cross-hybridization and more reliable detection of subtle rearrangement and deletion.

In assessment of the clinical significance of subtelomeric abnormalities, parental or family studies reveal a significant number are familial (i.e. are present in an unaffected parent or relative). Blake et al. [21] initially reported a list of subtelomeric abnormalities that were determined, in some cases, to be normal chromosome variants. The most frequent of these were 2q deletions (Fig. 30.2). Numerous studies have showed other telomeric regions to be duplicated, deleted or diminished in size in phenotypically normal individuals, some of which may or may not be normal variants. Examples of apparent deletions of 10q and 22q are shown in Fig. 30.3a, b. Although parental chromosome studies are recommended in every case to determine whether or not an abnormality is familial, not all cases may be so easily resolved. For example, apparent diminution in the size of a signal on 8p was consistently seen in a child with developmental delay. A similar but less dramatic discrepancy in signal size was seen in the 8 s of the child's phenotypically normal father (Fig. 30.3c). Hence, this case was not resolved by this FISH analysis but, in fact, some other form of molecular analysis would be required to determine if such a finding is real or artifact. Table 30.1 is extracted mainly from data recently published by Ravnan et al. on 11688 subtelomeric studies [18]. Of 357 subtelomeric abnormalities they

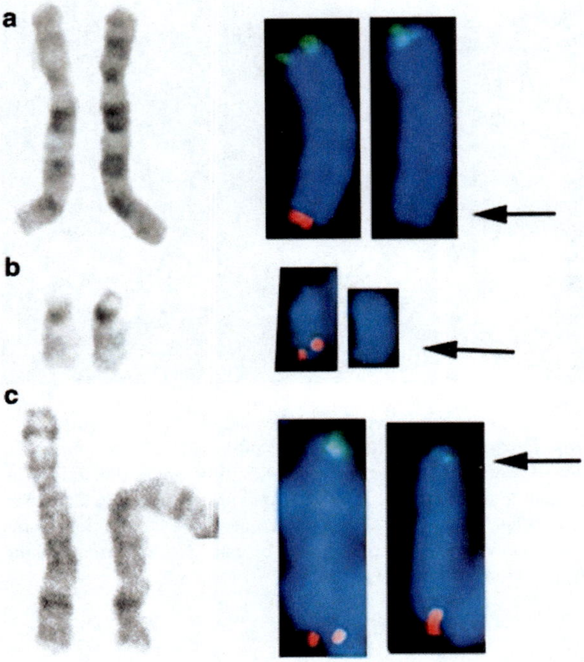

Fig. 30.3 Four cases showing what may or may not be normal variants at the telomeric ends of four different chromosomes (four different cases) showing loss of subtelomeric sequences (*arrows*) from 2q (**a**); from 8p (**b**); from 10q (**c**) and from 22q (**d**). In each case there appears to be no detectable loss of material by G-banding (pairs of chromosomes at *left*). Note that the loss from 8p (**b**) is not total and in fact an intermediate loss of sequences was seen in the father (Huang et al. unpublished). All of these cases were submitted for FISH analysis because of idiopathic mental and developmental retardation. In all such cases it is imperative that chromosomes of both parents also be studied by FISH, preferably using probes from the same source and confirmed, if possible, by additional molecular studies to demonstrate actual loss of genes

detected, 15.6% were familial or considered to be likely normal variants because the same variant was show to be familial in at least one other unrelated case. In 45% of their cases, neither parent was studied so that it was not known if the abnormality was familial or not.

Even if familial, it is not always clear to what extent epigenetic factors may play a role. Review of telomeric and subtelomeric structure by Reitmen [22, 23] reveals segmental duplications comprise 25% of the most distal 500 kb and 80% of the most distal 100 kb of DNA of each chromosome arm. Transcript families occurring in subtelomeric regions include non-coding pseudogenes and a variable subterminal transcript family corresponding to the 3′ end of genes coding for Wiscott-Aldrich syndrome (WASP) and Scar homolog (WASH) proteins. Other gene families em-

Table 30.1 Frequencies of neutral deletions and duplications by subtelomeric FISH

Chr	Ravan et al. [18] n = 11,688		Prior Publications [17,19–21] n = 2500		Total
	del	dup	del	dup	
2p				1	1
2q				20	20
3p	1		1	1	2
3q				1	1
4p	1	2			3
4q				1	1
6p				1	1
7q			1		1
8p		1	1		2
9p			27		27
10q	8		1		9
14q	1	1			2
16q	2				2
17p	1	1			2
17q				1	1
18p			1		1
20p	1				1
20q				1	1
Xp/Yp		4	5		9
Yq	3				3

bedded in duplicated or single copy subtelomeric DNA include immunoglobulin heavy chain, olfactory receptor, zinc-finger, and many gene families of unknown function. Epigenetic factors appear to play critical roles in regulating telomere replication, telomerase-dependent elongation, and telomeric recombination. Detailed maps of duplicons of individual chromosomes are available at: http:www.wistar.upenn.edu/Reithman.

In addition to parental or family studies, attempt is now frequently made to confirm subtle abnormalities suspected from subtelomeric FISH or other cytogenetic observation using CGH array technology. Some array CGH platforms incorporate "molecular rulers" to more accurately determine the size of duplications or deletion (Ballif et al. 2007).The array technologies (see Sect. E) in turn, however, frequently detect copy number variations (CNVs) that may not necessarily be found in data bases that list known variations. Figure 30.4, for example shows a subtelomeric deletion of 4q that is present in a woman who has a family history of recurrent pregnancy loss. Search of available microarray data bases (See Part IV), however, failed to identify other patients with similar sequence loss or history.

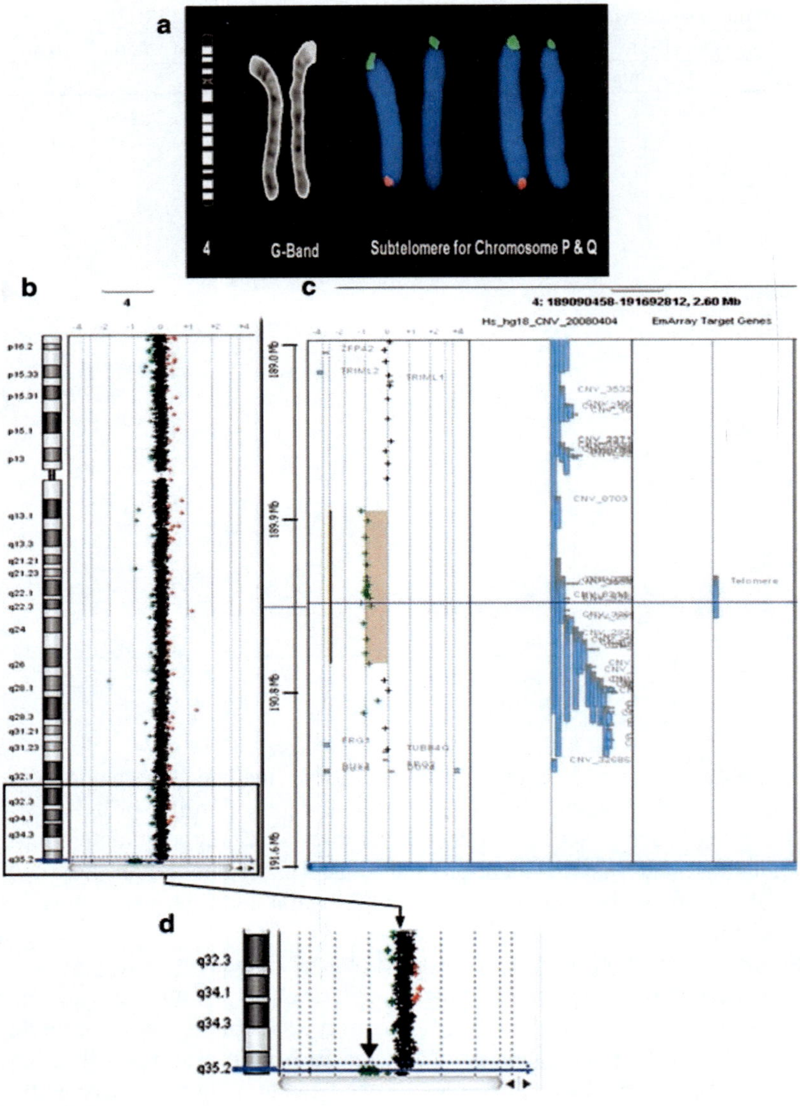

Fig. 30.4 Subtelomeric deletion of 4q in a woman with recurrent pregnancy loss, **a** not visible by G-baning (*left*) but clearly revealed by FISH using Vysis ToTelVysion subtelomeric panel, results of which showed no evidence of a balanced rearrangement (no signal for 4q elsewhere in the her chromosomes. **b–d** CGH EmArray Cyto6000 44 k olligonucleotide microarray on a commercial platform (Agilent Technologies), having an average probe spacing of 75 kb, confirmed a 0.8 Mb deletion in the 4q35.2 from bp 189,939,987 to bp190,706,472. Enlargment of area in rectangle **d** show deleted segment in *blue*. *Green* cluster (*Arrow*) is highlighted by tan bar **c** with overlapping clones containing target genes also shown in **c**. The woman has had two pregnancy losses and two in vitro fertilization failure. Her mother also had infertility and endometriosis but the woman and a twin brother were conceived on Clomid. The role, if any of the 4q deletion in this woman's infertility is uncertain. Information in available databases was unable to reveal cases with similar deletions or clinical findings, hence underscoring the difficulty in interpreting both subtelomeic FISH deletions and small deletions detected by array technology. (Vijay Tonk, Jee Hong Kyhm and Jennifer Phy, unpublished)

References

1. Buiting K, Dittrich B, Dworniczak B, Lerer I, Abeliovich D, Cottrell S, Temple IK, Harvey JF, Lich C, Grob S, Horsthemke B (1999) A 28-kb deletion spanning D15S63 (PW71) in five families: a rare neutral variant? Am J Hum Genet 65:1588–1594
2. Silverstein S, Lerer I, Buiting K, Abeliovich D (2001) The 28-kb deletion spanning D15S63 is a polymorphic variant in the Ashkenazi Jewish population. Am J Hum Genet 68:261–263
3. Pironon N, Peuchberty J, Roizes G (2010) Molecular and evolutionary characteristics of the fraction of human alpha satellite DNA associated with CENP-A at the centromeres of chromosomes 1, 5, 19 and 21. Genomics 11:195–213
4. Verlinsky Y, Ginsberg N, Chmura M, Freidine M, White M, Strom C, Kuliev A (1995) Cross-hybridization of the chromosome 13/21 alpha satellite DNA probe to chromosome 22 in the prenatal screening of common chromosomal aneuploidies by FISH. Prenat Diagn 15:831–834
5. Weir HU, Gray JW (1992) A degenerate alpha satellite probe, detecting a centromeric deletion on chromosome 21 in an apparently normal human male shows limitations of the use of satellite DNA probes for interphase ploidy analysis. Anal Cell Pathol 4:81–86
6. Lapidot-Lifson Y, Lebo RV, Flandermeyer R, Chung J-H, Golbus MS (1991) Rapid aneuploid diagnosis of high risk cases by fluorescence *in situ* hybridization. Am J Ob Gyn 174:886–890
7. Bossuyt PJ, Van Tienen MN, De Gruyter L, Smets V, Dumon J, Wauters JG (1995) Incidence of low-fluorescence alpha satellite region on chromosome 21 escaping detection of aneuploidy at interphase by FISH. Cytogenet Cell Genet 68:203–206
8. Tardy EP, Totu A (1997) Letters to the Editor. Cross-hybridization of the chromosome 13/21 alpha satellite DNA to chromosome 22 or a rare polymorphism. Prenat Diagn 17:487–490
9. Vissel B, Choo KH (1991) Four distinct alpha satellite subfamilies shared by human chromosomes 13, 14 and 21. Nucleic Acids Res 19:271–277
10. Choo KH, Earl E, Vissel B, Filby RG (1990) Identification of two distinct subfamilies of alpha satellite DNA that are highly specific for chromosome 15. Genomics 7:517–523
11. Stergianoff K, Gould CP, Waters JJ, Hulten MA (1993) A DA/DAPI positive human 14 heteromorphism defined by fluorescent *in situ* hybridization using chromosome 15-specific probes D15Z1 (satellite III) and p-TRA-25 (alphoid). Hereditas 119(2):105–110
12. Smeets DFCM, Merkx GFM, Hopman HM (1991) Frequent occurance of translocations of the short arm of chromosome 15 to other D-group chromosomes. Hum Genet 87:45–48
13. Shim SH, Pan A, Huang XL, Tonk VS, Varma SK, Milunsky JM, Wyandt HE (2003) FISH variants with D15Z1. J Associ Genet Technol. 29(4):146–151
14. Cockwell AE, Jacobs PA, Crolla JA (2007) Distribution of D15Z1 copy number polymorphism. Europ J Hum Genet 15:441–445
15. Earle E, Voullaire L, Hills L, Stlate H, Choo KHA (1992) Absence of satellite III DNA in the centromere and the proximal long-arm region of human chromosome 14: analysis of a 14p-variant. Cytogenet Cell Genet 61:78–80
16. Bonfatti A, Giunta C, Sensi A, Gruppioni R, Rubini M, Fontana F (1993) Heteromorphism of human chromosome 18 detected by fluorescent *in situ* hybridization. J Histochem 37:149–154
17. Jalal SM, Harwood A, Anderson M, Lorentz C, Law M, Lindor N, Karnes P, Kulharya A, Sekhon GS, Michels V (2000) Screening for subtle structural anomalies by use of subtelomeric specific probe set. Am J Hum Genet 67:A770
18. Ravnan JB, Teppenberg JH, Papenhausen P, Lamb AN, Hedrick J, Eash D, Ledbetter DH, Martin CL (2006) Subtelomeric FISH analysis of 11, 688 cases: an evaluation of the frequency and pattern of subtelomeric rearrangements in individuals with developmental disabilities. J Med Genet 43:478–489
19. Yu S, Baker E, Hinton L, Eyre HJ, Waters W, Higgins S, Sutherland GR, Hann E (2005) Frequency of truly cryptic subtelomere abnormalities—a study of 534 patients and literature review. Clin Genet 68:436–441

20. Knight SJ, Flint J (2000) Perfect endings: a review of subtelomeric probes and their use in clinical diagnosis. J Med Genet 57:401–409
21. Blake C, Kashork CD, Shaffer KG (2000) The promise and pitfalls of telomere region-specific probes. Am M Hum Genet 67:1356–1359
22. Reitman H (2008) Human telomere structure and biology. Annu Rev Genom Human Genet 9:1–19
23. Reitman H (2008) Human subtelomeric copy number variation. Cytogenet Genome Res. 123:244–252

Part III
Fragile Sites

Chapter 31
Fragile Sites

According to Sutherland and Hecht [1], fragility at specific sites on chromosomes were first described by Dekeban (1965) and by Lejeune (1968). Magenis et al. (1970) [2] used a heritable fragile site on chromosome 16 to map α-haptoglobin. The term "fragile site", however, is attributed to Frederick Hecht who, in referring to this site on chromosome 16, "wanted to convey the concept of transmissible points of chromosome fragility in the human genome" [1]. The clinically significant fragile site at Xq27.3 was initially reported by Lubs [3] as a familial marker X chromosome in four mentally retarded males and their normal mother. This marker was shown to be more than sporadically associated with a common form of X-linked male mental retardation in 1976–77 [4, 5]. However, it was not until Sutherland [6] showed that expression of the fragile X required culture in medium that was deficient in folic acid and thymidine that the fragile X in particular was accepted as a reproducible marker of the disease, and that other fragile sites became of interest as possible markers of disease. It was soon apparent that numerous other sites were sensitive to a variety of culture conditions [1]. These included: (1) folate-sensitive sites; (2) distamycin A-inducible sites; (3) BrdU–requiring sites; (4) common fragile sites (most strongly induced by aphidicolin, but weakly induced by conditions in the first three groups).

Fragile sites were initially described as having several essential features or properties [1]: (1) they were observed as a non-staining gap that varied in width for a particular site and usually involved both chromatids; (2) a particular site was always at exactly the same location within an individual or kindred; (3) it was inherited in a Mendelian co-dominant fashion; (4) it exhibited fragility under appropriate conditions of induction as acentric fragments, deleted chromosomes, triradial configurations, etc. There is a distinction between rare sites which can be demonstrated to be heritable and are seen in less than 5% of the population, and common sites that appear to be inducible in anyone. Both rare and common sites occur at specific locations on chromosomes; however, common fragile sites are ubiquitous to all individuals (often expressed homozygously) under condition of replicative stress. Most common sites are induced *in vitro* in normal human white blood cells by aphidicolon or 5-azacytidine [7, 8] whereas, most rare fragile sites are expressed in folic-acid (enhanced by methotrexate) or thymidine deficient (enhanced by FudR)

H. E. Wyandt, V. S. Tonk, *Human Chromosome Variation: Heteromorphism and Polymorphism,* DOI 10.1007/978-94-007-0896-9_31,

media, with some being induced by 5-bromodeoxyuridine (5-BdU) or distamycin A [9–11]. Numerous reports have shown that most common fragile sites appear to coincide with non-random sites of breakage in chromosome rearrangements [1, 8–10, 13, 14]. A high proportion is recurrent in cancer rearrangements [11, 12].

To date, approximately 90 common fragile sites and 30 rare fragile sites are recognized in the human genome data base, identified by official HUGO nomenclature [9–13]. However, a recent report by Mrasek et al. [14] suggests as many as 230 aphidicolin-inducible fragile sites, 52 of which were previously reported but not given HUGO identities, and 61 of which have not been previously reported.

At least nine rare fragile sites (seven folate-sensitive and two non-folate sensitive) and thirteen aphidicolin inducible (common fragile sites) have been molecularly characterized [13]. Molecular characterization reveals these sites to have features that predispose them to form hairpin loops, DNA slippage, tetrahelical structures and DNA flexion, all of which can contribute to disturbance of DNA replication and/or chromatin assembly, and render them highly recombinogenic in translocations, deletions, amplifications and as sites for exogenous DNA insertion [9]. Their inherent instability involves a number of mechanisms, which have in common the partial disruption or inhibition of DNA replication without arrest of the cell cycle [7, 9, 15, 16].

Several of the most common sites tend to be large genomic regions, a megabase in size or larger, usually AT-rich and sometimes late-replicating [15]. Several have been shown to be the sites of large cancer genes, the most notable of which are FHIT at 3p14.2, WWOX at 16q23.2, Parkin at 6q26 and GRID2 at 4q22 [11, 12].

Of rare fragile sites, FRAXA, causing fragile X syndrome, and FRAXE, associated with X-linked non-specific mental retardation, are clinically the most significant sites [17]. No other sites have been clearly associated with inherited disease. Some rare fragile sites appear to correlate with breakpoints giving rise to deletions and rearrangements. FRA11B is of interest in that it is possibly implicated in the generation of deletions in a proportion of patients with Jabobsen syndrome; however the majority of deletion breakpoints in this syndrome map away from the fragile site [15]. The most common rare site, FRA16B, has also been implicated in a number of reports of deletion at 16q22 in various forms of cancer and leukemia, and was of interest in a recent report of hereditary neutropenia in a mother and daughter [18].

Table 31.1 lists both rare (in bold) and common sites, first by chromosome number and then in band order on the chromosome, starting at the terminal end of the short arm and ending at the terminal end of the long arm. The official HUGO identity of each site is given as well as its ISCN identity by band location. Some sites that are induced by more than one agent or condition are listed more than once. Some are listed as a common site by one type of induction and as a rare site by another. Also listed are genes and cancer genes associated with sites where these are know.

Figs. 31.1–31.8 shows a number of aphidicolin-induced sites listed in Table 31.1. Some aphidicolin-induced sites that have not been given official HUGO designations. Others are among those reported by Mrasek et al. [14] and others.

Table 31.1 List of common and rare fragile sites currently recognized by HGCN*

	Fragile site I.D.*	Locus		Induced by	Associated gene(s)	Comments
1	FRA1A	1p36**				Possibly associated with cancer, site was early reported to be within region frequently deleted in neuroblastoma [19-20]
2	FRA1B	1p32**			DAB1	Disabled-1: This gene (similar to disabled-1 in mice) and the protein encoded by this gene is thought to be a signal transducer that interacts with protein kinase pathways to regulate neuronal migration and lamination in the developing cerebral cortex. DAB1 expression is increased in several cancers, especially brain and endometrial [21].
3	FRA1C	1p31.2				
4	FRA1L	1p31				
5	FRA1D	1p22**				
6	FRA1E	1p21.2				FRA1E common fragile site breaks map within a 370 kilobase pair region and disrupt the dihydropyrimidine dehydrogenase gene (DPYD) [22].
7	**FRA1M**	**1p21.3**	**rare**	**folate-free**		Heritable site was found in a 32 yo woman and in her normal 75 yo mother. The woman had several miscarriages, one a fetus with der(9)t(1;9), andher partner had a balanced t(1:9). Previous wife also had had three miscarriages. Fragile site, therefore, was likely benign [23]
8	FRA1J	1q12		5-aza		
9	FRA1F	1q21				
10	FRA1G	1q25**				
11	FRA1K	1q31				
12	FRA1H	1q42.1		5-aza		
13	FRA1I	1q44				
14	FRA2C	2p24.2**				
15	FRA2D	2p16.2				
16	FRA2E	2p13**				
17	**FRA2A**	**2q11.2****	**rare**	**folate-free**		

Table 31.1 (continued)

	Fragile site I.D.*	Locus		Induced by	Associated gene(s)	Comments
18	**FRA2B**	**2q13****	**rare**	**folate-free**		Site shown to be distal to an inverted interstitial telommeric/subtelomeric array thought to result from an ancestral fusion in primates. Fragile site, therefore, appears to be distinct from fusion site [24]
19	FRA2F	2q21.3			LRP1B	Low-density lipoprotein receptor-related protein maps to 2q21.2, a region found to be deleted in some aggressive cancers [25] and is also an integration site for HPV18 in cervical cancer [11]
20	**FRA2K**	**2q22.3**	**rare**	**folate-free**		
21	FRA2G	2q31**			IGRP, RDHL, LRP2	The large 450 kb fragile site is located within a 1Mb region that spans at least 7genes at least one of which was noted to be a candidate tumor suppressor gene in a number of cancers. Several regions of high flexibility were found within the fragile site that potentially contribute to fragility [26], listed as LRP2 by Glover et al [8, 16]. LRP1B is reported to be an integration site for HPV16 [11]
22	FRA2H	2q32.1				
23	FRA2I	2q33				
24	FRA2J	2q37.3				
25	FRA3A	3p24.2**				
26	FRA3B	3p14.2**			FHIT	Most frequently expressed CFS, this gene, a member of the histidine triad gene family, encodes a diadenosine triphosphate hydrolase involved in purine metabolism. The gene includes the common fragile site FRA3B on chromosome 3, where carcinogen-induced damage can lead to translocations and aberrant transcripts, found in about half of all esophageal, stomach, and colon carcinomas. Alternatively, spliced transcript variants have been found for this gene [11-12].
27	FRA3D	3q25				
28	FRA3C	3q27**				
29	FRA4A	4p16.1				
30	FRA4D	4p15				
31	FRA4B	4q12		5-BrdU		

Table 31.1 (continued)

	Fragile site I.D.*	Locus	Induced by	Associated gene(s)	Comments
32	FRA4F	4q22		K1AA1680, GRID2	GRDD2, coding for the ionotropic glutamine receptor delta 2 gene, has recently been reported as a hotspot for mutation and translocation in human and mouse [8,11-12,16]
33	FRA4E	4q27	Unclass.		
34	FRA4C	4q31.1			
35	FRA5E	5p14			
36	FRA5A	5p13	5-BrdU		
37	FRA5B	5q15			
38	FRA5D	5q15	5-BrdU		
39	FRA5F	5q21			
40	FRA5C	5q31.1**			
41	**FRA5G**	**5q35****	**rare folate-free**		Site ascertained in family with history of male mental retardation. Site was present in a retarded male but absent in his mother, a normal brother, two mentally retarded maternal uncles and his nomal maternal grandfather [27].
42	FRA6B	6p25.1			
43	**FRA6A**	**6p23****	**rare folate-free**		
44	FRA6C	6p22.2			
45	FRA6D	6q13	5-BrdU		
	FRA6G	6q15			
47	FRA6F	6q21**		REV3L, DIF13	A 1200 kb region which transcribes 19 known genes, 10 of which span two hotspots of breakage spanning 200 kb. Deletion breakpoints within site are common in several types of leukemias and solid tumors and include REV3L, DIF13, dj1112D6.1, C6UAS, C60RFA-6, FYN, H3F3Am FKHRL1, dj142L7.3, and LAMA4 [28]. Site is also listed by GLover et al [8,16] as a cancer prone site.
48	FRA6E	6q26**		MAP3K4, LPA, PARKIN	Chromosome break at FRA6E is associated with breast cancer. FRA6E shares many similarities with FRA3B (3p14.2) and FRA16D (16q23.2) in representing a large region of genomic instability and containing an extremely large gene that may play a role in the development of ovarian and many other cancers [11-12, 16,29]

Table 31.1 (continued)

	Fragile site I.D.*	Locus	Induced by	Associated gene(s)	Comments
49	FRA7B	7p22			
50	FRA7C	7p14.2			
51	FRA7D	7p13**			
52	FRA7J	7p11**			
53	**FRA7A**	**7p11.2**	**rare folate-free**		
55	FRA7E	7q21.11		LEP	This gene encodes a protein that is secreted by white fat cells and plays a role in regulating body weight. It acts through a leptin receptor and is part of a signaling pathway that can inhibit food intake and regulate energy expenditure. It also regulates inmune and inflammatory responses, hematopoiesis angiogenesis and wound healing. Mutations can cause morbid obesity with hypogonadism and has been linked to type 2 diabetes [8,16].
56	FRA7F	7q22**			
57	FRA7G	7q31.2		TESS; TESS-2; MGC1146; DKFZp586B2022; TES; CAV1, CAV2, TESTIN, MET; D7S486	Cancer-associated chromosomal changes often involve regions containing fragile sites. This gene maps to a commom fragile site on chromosome 7q31.2 designated FRA7G. This gene is similar to mouse Testin, a testosterone-responsive gene encoding a Sertoli cell secretory protein containing three LIM domains. LIM domains are double zinc-finger motifs that mediate protein-protein interactions between transcription factors, cytoskeletal proteins and signaling proteins. Multiple protein isoforms are encoded by transcript variants of this gene [8,16].
58	FRA7H	7q32.3 **			
59	FRA7I	7q36			
60	**FRA8F**	**8q13**	**rare unclass**		Two failies of highly repetitive elements at FRA8F that may confer fragility and have some regulatory function flank the 3' end of corticotropin releasing hormone (hCRH) [30]
61	FRA8B	8q22.1			
62	**FRA8A**	**8q22.3****	**rare folate-free**		
63	FRA8C	8q24.1			

Table 31.1 (continued)

	Fragile site I.D.*	Locus		Induced by	Associated gene(s)	Comments
64	**FRA8E**	**8q24.1**	**rare**	**Dist-A**		Site appears to be adjacent to region of hereditary multiple exostoses and Langer-Giedion syndrome deletion endpoints, but proximal to MYC gene in 8q24.12-q24.13 [31-33]
65	FRA8D	8q24.3				
66	FRA9D	9p22.1				
67	**FRA9A**	**9p21****	**rare**	**folate-free**		
68	FRA9C	9p21		5-BrdU		
69	FRA9F	9q12		5-Aza		
70	**FRA9B**	**9q32****	**rare**	**folate-free**		Sutherland (1982) found 1 example of 9q32 fragile site in a population study [34]
71	FRA9E	9q32-33.1			PAPPA, ROD1, KLFA	PAPPA (pregnancy-associated plasma protein is showing frequent loss of expression in ovarian cancer lines and LOH in ovarian tumors is located in the distal end of FRA9E. FRA9E extends 9Mb, contans 16 genes and is the largest CFS characterized to date [35]
72	FRA10G	10q11.2				
73	FRA10C	10q21		5-BrdU		
74	FRA10D	10q22.1				
75	**FRA10A**	**10q23.3****	**rare**	**folate-free**	**FRA10AC1**	FRA10C1, transcribed centromeric of FRA10A, is a novel gene consisting of 19 exons, whose major translcript of ~1450 nt is a highly conserved protein of unknown function. Expansions >200 CCG repeats in expressed FRA10A are methylated [13,36]. Several reports have described clones in amniotic fluid that have deletion at or close to 10q23.3, but with normal outcomes, including a case with 100% exprssion expression (3/22 cells with deletion) in amniotic fluid for a phenytoin-exposed fetus, which was subsequently terminated and showed no external anomalies on autopsy [37].
76	**FRA10B**	**10q25.2****	**rare**	**5-BrdU**		
77	FRA10E	10q25.2				
78	FRA10F	10q26.1				
79	FRA11C	11p15.1				

Table 31.1 (continued)

	Fragile site I.D.*	Locus		Induced by	Associated gene(s)	Comments
80	**FRA11I**	**11p15.1**	**rare**	**Dist-A**		
81	FRA11D	11p14.2				
82	FRA11E	11p13**				
83	FRA11H	11q13				
84	**FRA11A**	**11q13.3****	**rare**	**folate-free**	**C11 or f8C**	A relationship between cytogenetic expression of the fragile site and the mental handicap seems unlikely, as FRA11A was found in a mentally retarded patient as well as in phenotypically normal carriers from the same family [38].
85	FRA11F	11q14.2				Gene amplification involving chromosomal band 11q13 in oral squamous-cell carcinomas may be initiated by breakage at FRA11F [39].
86	**FRA11B**	**11q23.3****	**rare**	**folate-free**	**CBL2**	One of seven folate sensitive sites that have been fully characterized, FRA11B has been suggested to be implicated in deletions in patients with Jacobsen syndrome which maps close to the fragile site. The majority of deletions in Jacobsen syndrome, however, map away from FRAllB [13].
87	FRA11G	11q23.3				Localized to a 4.5 Mb region, 0.8 Mb proximal to genomic region affected by expression of FRA11B [40].
88	**FRA12A**	**12q13.1****	**rare**	**folate-free**	**DIP2B**	FRA12A is a folate-sensitive chromosomal fragile site prone to breakage. No consistent phenotype has been observed with FRA12A, and it can be inherited without phenotypic effect (Berg et al., 2000 [PubMed 10955484]). However, mental retardation with or without other anomalies has been described in patients with over 40% of cells expressing FRA12A [41].
89	FRA12B	12q21.3				
90	**FRA12C**	**12q24**	**rare**	**5-BrdU**		
91	FRA12E	12q24				
92	**FRA12D**	**12q24.13**	**rare**	**folate-free**		
93	FRA13A	13q13,2			NBEA	The neurobeachin gene (NBEA) spans the common fragile site FRA13a, with breaks in the site occuring within a 650 kb region within the gene. NBEA is a recurrent target for interstitial deletions at 13q13 in MGUS and multiple myeloma [11-12, 42-43]

Table 31.1 (continued)

	Fragile site I.D.*	Locus	Induced by	Associated gene(s)	Comments
94	FRA13B	13q21.2	5-BrdU		
95	FRA13C	13q21.2			
96	FRA13D	13q32			
97	FRA14B	14q23			
98	FRA14C	14q24.1**			
99	FRA15A	15q22		RORA	This gene encodes a protein that is a member of the NR1 subfamily of nuclear hormone receptors and can bind to hormone response elements upstream of genes that enhance their expression [11-12].
100	**FRA16A**	**16p13.11** **rare**	**folate-free**		
101	**FRA16E**	**16p12.1**** **rare**	**Dist-A**		
102	**FRA16B**	**16q22.1**** **rare**	**Dist-A**		
103	FRA16C	16q22.1			
104	FRA16D	16q23.2**		WWOX	A model is supported in which FRA16D common fragile sites are sequences that may initiate replication in early-mid S phase but are slow to complete replication. The chromosomal breaks and gaps observed in metaphase cells result from unreplicated DNA [8,11-12, 44]
105	**FRA17A**	**17p12**** **rare**	**Dist-A**		
106	FRA17B	17q23.1			
107	FRA18A	18q12.2			
108	FRA18B	18q21.3			
109	FRA18C	18q22.1-q22.2			Cloned breakpoint found to disrupt the DOK6 gene and to be followed by a telomeric repeat motif. Site correscponded to breakpoint in a girl with Beckwith-Wiedemann syndrome who had *in vivo* truncation of 18q22-18qter, and with fragile site in the father [45].
110	FRA19A	19q13	5-Aza		
111	**FRA19B**	**19p13** **rare**	**folate-free**		
112	FRA20B	20p12.2			

Table 31.1 (continued)

	Fragile site I.D.*	Locus		Induced by	Associated gene(s)	Comments
113	**FRA20A**	**20p11.23****	**rare**	**folate-free**		
114	**FRA22A**	**22q12.2**	**rare**	**folate-free**		
115	FRA22B	22q12.2**				
116	FRAXB	Xp22.31**			STS	Deletion or mutation of STS causes steroid sulfatase deficiency [16]
117	FRAXC	Xq22.1**				
118	FRAXD	Xq27.2				No CGG repeat, no disease association
119	**FRAXA**	**Xq27.3****	**rare**	**folate-free**	**FMR1**	Fragile X syndrome associated with CGG repeat [18]
120	**FRAXE**	**Xq28**	**rare**	**folate-free**	**FMR2**	Mild X-linked mr associated with CGG repeat[18]
121	**FRAXF**	**Xq28**	**rare**	**folate-free**	**FAM1A**	First described by Hirst et al (1993) as one of three expanded, hypermethylated and unstable CGG repeats with CpG islands. Ritchie et al (1994) showed that expansion involved a (GCCGTC)n(GCC)n compound array consisting of 12-26 repeats in normal individuals. Shaw et al (2002) identified FAM11A, a 350 amino acid protein, as distal to encompassing the FRAXF CpG island. Expansion of the CGG site resulted in methylation of CpG island and transcriptional silencing of FAM11A [18]

Loci highlighted in yellow in the above table are shown in Figures D1-D8 that follow. Rare fragiles are shown in bold

* HGNC ID: Sites in the above table are officially recognized by the listed fragile site I.D. by the Human Genome Nomenclature Committee (HGNC): See website, http://www.genenames.org/index.html

** Fragile sites listed in Sutherland and Hecht (1985) [6]: Of 41 listed sites, some 14 were listed as corresponding to cancer breakpoints; others were in the vicinity of a breakpoint or on a chromosome frequently involved in cancer. The distinction between "common" and "rare" was uncertain for a number of sites at that time, as was the precise locus for several sites. Despite the imprecision, the association with cancer genes was especially prescient, as revealed by more recent molecular investigations cited in the above table.

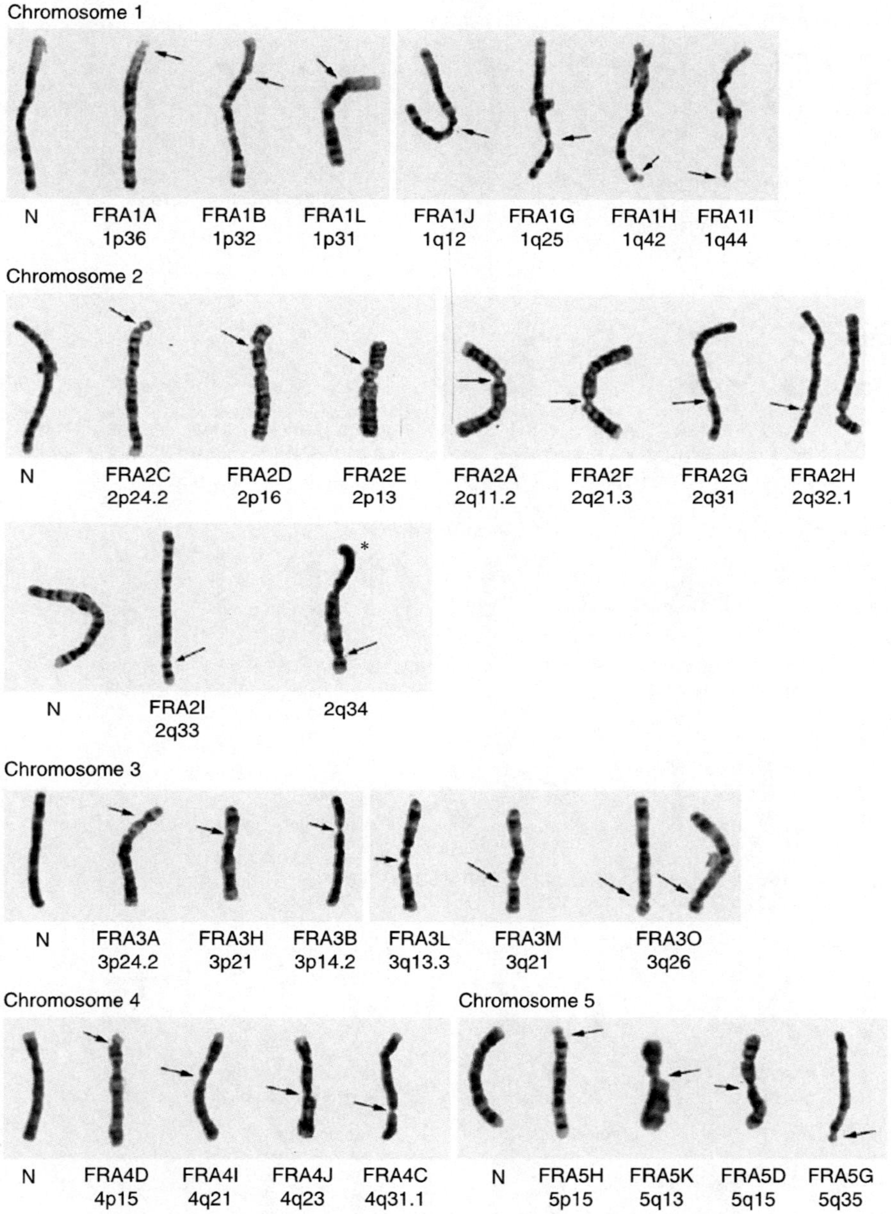

Fig. 31.1–31.8 Examples of common fragile sites for chromosomes 1, 2 and 3 observed in cultures treated with aphidicolin. Sites (*) shown as breakpoints without a HUGO designation are presumably random breaks or gaps that have not been established as common fragile sites. (Prepared and contributed by Caro Gibson and Eun-Jung Lee, Texas Tech University Health Sciences Center, Lubbock , TX)

Fig. 31.1–31.8 (continued)

References

1. Giraud F, Aymé S, Mattei JF, Mattei MG (1976) Constitutional chromosome breakage. Hum Genet 34:125–136
2. Harvey J, Judge C, Wiener S (1977) Familial X-linked mental retardation with an X chromosome abnormality. J Med Genet 20:280–285
3. Lubs HA (1969) A marker X chromosome. Am J Hum Genet 21:231–244
4. Sutherland GR (1977) Fragile sites on human chromosomes: demonstration of their dependence on the type of tissue culture medium. Science 197:265–266
5. Sutherland GR (1979) Heritable fragile sites on human chromosomes. I. Factors affecting expression in lymphocyte culture. Am J Hum Genet 31:125–135
6. Hecht F, Sutherland GR (1985) Fragile sites on human chromosomes. Oxford University Press, New York
7. Glover TW, Berger C, Coyle J, Echo B (1984) DNA polymerase α inhibition by amphidicolin induces gaps and breaks at common fragile sites in human chromosomes. Hum Genet 67:136–142
8. Glover TW (2006) Common fragile sites. Cancer Lett 232(1):4–12. Epub 2005 Oct 17. Review
9. Schwartz M, Zlotorynski E, Kerem B (2006) The molecular basis of common and rare fragile sites. Cancer Lett 231(1):13–26. Epub 2005)ct 19. Review
10. DeBacker K, Kooby F (2007) Fragile sites and human disease. Hum Mol Genet 16(2):R150-–R158
11. Smith DI, McAvoy S, Zhu Y, Perez DS (2007) Large common fragile site genes and cancer. Semin Cancer Biol 17:31–41
12. Smith DI, Zhu Y, McAvoy S, Kuhn R (2006) Common fragile sites, extremely large genes, neural development and cancer. Cancer Lett 232:48–57
13. Lukusa T, Fryns JP (2008) Human chromosome fragility. Biochimica Biopysica Acta 1779:3–16
14. Mrasek K, Schoder C, Techmann A-C, Behr K, Franze B, Wilhelm K, Blaurock N, Claussen U, Liehr T, Weise A (2010) Global screening and extended nomenclature for 230 aphidicolin-inducible fragile sites, including 61 yet urepoted ones. Interntl J Oncol 36:929–940
15. Picchiorri F, Ishii H, Okumura H, Trapasso F, Wang Y, Heubner K (2008) Molecular parameters of genome instability: roles of fragile genes at common fragile sites. J Cell Biochem 104:1525–1533
16. Glover TW, Arit MF, Casper AM, Durkin SG (2005) Mechanisms of common fragile site instability. Hum Mol Genet 14:197–205
17. Glasser L, Meloni-Ehrig A, Josesph P, Mendiola J (2006) Benign chronic neutropenia with abnormalities ivolving 16q22, affecting mother and daughter. Am J Hemat 81:262–270
18. Sutherland GR, Brown WT, Hagerman R, Jenkins E, Lubs H, Mandel J-L, Neslon D, Neri G, Partington MW, Richards RI, Stevenson R, Turner G (1994) Sixth international workshop on the fragile X and X-linked mental retardation. Am J Med Genet 51:281–293
19. Haag HM, Soucup SW, Neely JE (1981) Chromosome analysis of a human neuroblastoma. Cancer Res 41:2595–2599
20. Sutherland GR, Hecht F (1985) Fragile Sites on Human Chromosomes. Oxford Monographs on Medical Genetics No. 13. Oxford University Press, New York, pp 207–226
21. MAvoy S, Zhu Y, Perez DS, James CD, Smith DI (2008) Disabled-1 is a large common fragile site gene, inactivated in multiple cancers. Genes Chromosomes Cancer 47(2):165–174
22. Hormozian F, Schmitt JG, Sagulenko E, Schwab M, Savelyeva L (2007) FRA1E common fragile site breaks map within a 370kilobase pair region and disrupt the dihydropyrimidine dehydrogenase gene (DPYD). Cancer Lett 246(1–2):82–91
23. Baker E, Sutherland GR (1991) A new folate sensitive fragile site at 1p21.3. J Med Genet 28:356–357

24. IJdo JW, Baldini A, Wells RA, Ward DC, Reeders ST (1992) FRA2B is distinct from inverted telomere repeat arrays at 2q13. Genomics 12:833–835
25. Liu CX, Musco S, Lisitsina NM, Yaklichkin SY, Lisistsyn NA (2000) Genomic organization of a new candidate tumor suppressor gene, LRP1B. Genomics 69:271–274
26. Limongi MZ, Pellicia F, Rocchi A (2003) Characterization of the human common fragile site FRA2G. Genomics 81:93–97
27. Howell RT, McDermott A, Evans JL (1990) A new apparently folate sensitive fragile site, 5q35. J Med Genet 27:527–528
28. Morelli C, Karayianni E, Magnanini C, Mungal AJ, Thorland E, Negrini M, Smith DL, Barbanti-Brodano G (2002) Cloning and characterization of the common fragile site FRA6F harboring a replicative senescence gene and frequently deleted in human tumors. Oncogene 21(47):7266–7276
29. Denison SR, Callahan G, Becker NA, Phillips LA, Smith DI (2003) Characterization of FRA6E and its potential role in autosomal recessive juvenile parkinsonism and ovarian cancer. Genes Chromosomes Cancer 38:40–52
30. Vamvakopoulos NC, Chrousos GP (1993) Structural organization of the 5' flanking region of the human corticotrophin releasing hormone gene. DNA Seq 4:197–206
31. Hori T, Seki N, Ohira M, Saito T, Yamauchi M, Sagaara M, Hayashi A, Tsuji S, Ito H, Imai T (1998) A distamycin A-inducible fragile site, FRA8E located I the region of the hereditary multiple exostoses gene, is not involved in HPV16 dNA integration and amplification. Cancer Genet Cytogenet 101:24–34
32. Hill A, Harada Y, Takahashi E, Hou J, Wagner MJ, Wells DE (1997) Assignment of fragile site 8E (FRA8E) to human chromosome band 8q24.11 adjacent to the hereditary multiple exostoses 1 geneand two overlapping Langer-Giedion syndrome deletion endpoints. Cytogenet Cell Genet 78:56–57
33. Takahashi E, Hori T, O'Connell P, Leppert M, White R (1991) Mapping the MYC gene to band 8q24.12-q24.13 by R-banding and distal to fra(8)(q24,11), FRA8E, by fluorescence *in situ* hybridization. Cytogenet Cell Genet 57:109–111
34. Sutherland GR (1982). Heritable fragile sites on human chromosomes. IX. Population cytogenetics and segregation analysis of the BrdU-requiring fragile site at 10q25. Am J Hum Genet 34(5):753–756
35. Callahan G, Dennison SR, Phillips LA, Schridhar V, Smith DI (2003) Characterization of the common fragile site FRA9E and its potential role in ovarian cancer. Oncogene 22:590–601
36. Sarafidou T, Kahl C, Martinez-Garay I, Mangelsdorf M, Gesk S, Baker E, Kokkinaki M, Talley P, Maltby EL, French L, Harder L, Hinzmann B, Nobile C, Richkind K, Finnis M, Deloukaas P, Sutherland GR, Kutsche K, Moschonas NK, Siebert R, Gécz J (2004) Folate sensitive fragile site FRA10A is due to an expansion of a CGG repeat in a novel gene, FRA10AC1, endcoding a nuclear protwin. Geneomics 84:69–81
37. Morel CF, Duncan AMV, Désilets V (2005) A fragile site at 10q23 (FRA10A) in a phenytoin-exposed fetus: a case report and review of the literature. Prenat Diagn 25:318–321
38. Debacker K, Winnepenninckx B, Longman C, Colgan J, Tolmie J, Murray R, van Luijk R, Scheers S, Fitzpatrick D, Kooy F. (2007). The molecular basis of the folate-sensitive site FRA11A at 11q13. Cytogenet Genome Res 119(1–2):9–14
39. Reshmi SC, Huang X, Schoppy DW, Black RC, Saunders WS, Smith DI, Gollin SM (2007) Relationship between FRA11F and 11q13 gene amplification in oral cancer. Genes Chromosomes Cancer 46(2):143–154
40. Fechter A, Buettel I, Kuehnel e, Savelyeva L, Schwab M (2007) Commmon fragile site FRA11G and Rare fragile site FRA11B at 11q23.3 encompass distince genomic regions. Genes Chromoosomes Cancer 46:98–106
41. Winnepenninckx B, Debacker K, Ramsay J, Smeets D, Smits A, FitzPatrick DR, Kooy RF (2007) CGG repeat expansion in the DIP2B gene is associated with the fragile site FRA12A on chromosome 12q13.1. Am J Hum Genet 80(2):221–231
42. Savelyeva L, Sagulenko E, Schmidt JG, Schwab M (2006) The neurobeachin gene spans the common fragile site FRA13A. Hum Genet 118:551–558

43. O'Neal J, Gao F, Hassan A, Monahan R, Barrios S, Kilimann MW, Lee I, Chng WJ, Vij R, Tomasson MH (2009). Neurobeachin is a target of recurrent interstitial deletions at 13q13 in patients with MGUS and multiple myeloma. Exp Hematol 37(2):234–244. (Erratum in: Exp Hematol. 37(4):532. Kilimann, Manfred W).
44. Palakodeti A, Han Y, Jiang Y, Le Beau MM (2004) The role of late/slow replication of the FRA16D in common fragile site induction. Genes Chromosomes Cancer 39:71–76
45. Debacker K, Winnepenninckx B, Ben-Porat N, FitzPatrick d, Van Luijk R, Scheers S, Kerem B, Frank Kooy R (2007). FRA18C: a new aphidicolin-inducible fragile site on chromosome 18q22, possibly associated with in vivo chromosome breakage. J Med Genet 44:347–352

Part IV
Copy Number Variants

Chapter 32
Copy Number Variants

32.1 Introduction

The progression of cytogenetics from banded chromosomes to DNA segment dosage as detected by comparative genomic hybridization-microarray analysis (aCGH, also abbreviated as CMA) has greatly increased the frequency of positive findings and difficulties of interpretation. The technique of aCGH (Fig. 32.1) involves labeling patient and control DNA with different fluorochromes, hybridizing them to 40,000–1 million DNA segments arrayed on a glass slide or "DNA chip," and comparing the extent of patient and control DNA hybridization signal amplitudes for each segment [1–3]. The DNA segments serving as probes are chosen to represent the entire genome extending through chromosomes 1–22, X, and Y, allowing graphing of hybridization signals according to chromosome band and base pair coordinates (see examples in patient discussions below). Departures of hybridization ratios from equivalency indicate duplication (patient signal greater than control) or deletion (patient signal less than control), and the number of contiguous DNA segments showing signal alterations define the extent of duplication or deletion by reference to their coordinates in the human genome sequence.

Since aCGH only detects differences in DNA segment dosage, it will not register balanced chromosome rearrangements such as balanced reciprocal translocations or inversions. The technique is being used to detect subtle deletion/duplication surrounding the breakpoints of apparently balanced translocations, but its omission of repetitive DNA segments will not register deletions attending the formation of Robertsonian translocations (e. g., 45,XY,der(14;21)(q10;q10)—thus, aCGH cannot replace routine chromosome analysis in distinguishing trisomy from translocation 21, etc. Populating the array grid with segments containing single nucleotide polymorphisms (SNPs) can discriminate among parental alleles (e. g. paternal allele –G– versus maternal allele –A–), and thus detect some instances of uniparental disomy as well as parental consanguinity (increase in homozygosity). Use of oli-

Written by Drs. Golder N. Wilson and Vijay Tonk; edited by Drs. Vijay Tonk and Herman Wyandt.

H. E. Wyandt, V. S. Tonk, *Human Chromosome Variation: Heteromorphism and Polymorphism,* DOI 10.1007/978-94-007-0896-9_32,

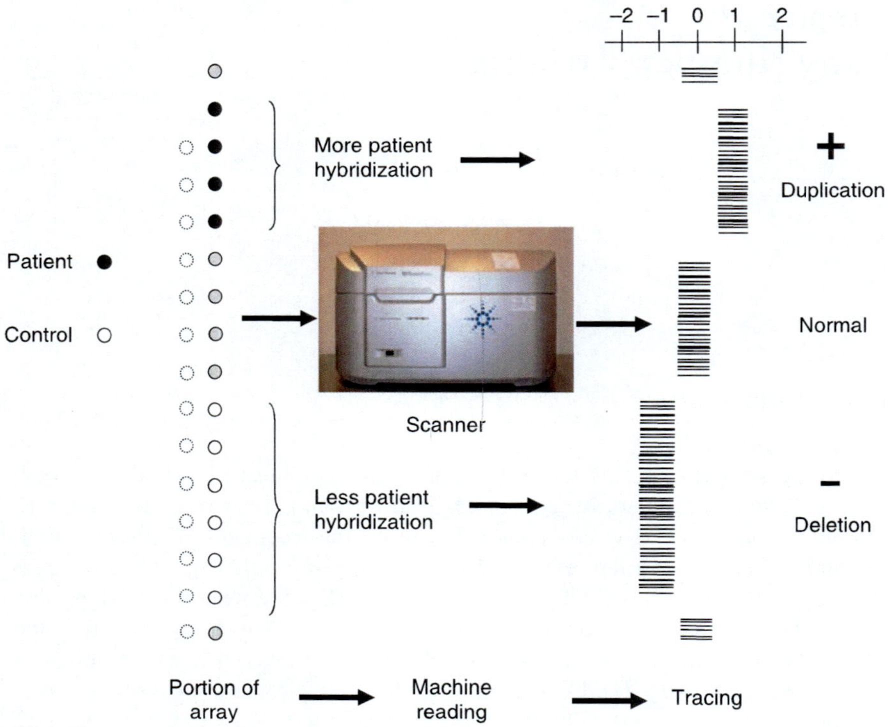

Fig. 32.1 Patient and control DNA is isolated (routine testing uses leukocytes from blood), labeled with different fluorochromes, and hybridized to an array of DNA segments chosen to represent the entire genome (*left*). Hybridization signals are analyzed by machine (*middle*) and plotted to yield a profile of comparative hybridization (patient versus control) for each chromosome (*right*)

gonucleotide versus SNP arrays, the number of DNA segments placed on the chip (determining the resolution of duplication/deletion segment length), and the selection of DNA segment coverage to target particular disease-associated regions are areas of aCGH technology that are evolving.

The duplication and deletion changes detected by aCGH are termed 'copy number variants' or CNVs, presaged by the chromosome 17 segmental duplication associated with Charcot-Marie-Tooth disease [4] and now recognized to be present in most individuals [5]. The detection of CNVs thus extends appreciation of dosage variation from large segmental aneuploidies or expansion/contraction of repeats to micro-changes in single-copy DNA. The term 'variant' is used for CNVs because their population frequencies are not known; they thus cannot be described as polymorphisms where variant frequencies are defined as above 1% but are required to have sizes greater than the 1 kilobase (kb) limit for transposable elements [5]. Most are thought to arise by recombination between non-homologous, low-copy repeats (Alu or satellite DNA sequences) that are distributed over most genomic regions. A chromosome region of structure 1—2—3—4 where the dashes represent repeats and the numbers represent single copy segments can thus undergo non-homologous cross-over between repeats, exemplified by cross-over between repeat 1–2 and repeat 3–4 to produce the structure 1–4 that has excised single copy segments 2 and 3 [6, 7]. Most CNVs affect one of the paired chromosomes although homozygous deletions have been reported [5].

Use of aCGH for diagnostic testing has increased the yield of positive findings in children with intellectual disability some 5–6 fold—from 3% for routine chromosome analysis to 10–12% in some studies [8] or 15–20% in a recent review [9]. The higher yield has led some to recommend aCGH as the first-tier test for children with ID but without 'obvious' chromosome disorders like Down syndrome, although situations where translocations are common (as in couples with recurrent miscarriage) will require use of routine karyotyping [9]. However, the frequency of CNVs in normal individuals makes interpretation of positive CNV findings difficult. Guidelines defining a CNV as significant (related to the patient's condition) include a size greater than 0.5 Mb, absence in normal parents, duplication rather than the more common deletion, multiplicity of known genes within the aneuploid segment, and absence from expanding databases of benign CNVs [10]. We will now present several cases that illustrate the types of aCGH findings and approaches to their interpretation.

32.2 Case Discussions

32.2.1 Cases Where the Interpretation of Clinical Significance Is Clear and the Diagnosis Provides Good Prognostic Information

Case 1: A 6-year-old female (Fig. 32.2a) has had mild motor delay and now shows poor speech articulation. Her speech therapist is concerned about palatal dysfunction. Clinical genetic evaluation shows a slightly abnormal facial appearance with prominent nose that suggests a diagnosis of the DiGeorge/velocardiofacial spectrum. Routine chromosome analysis is normal but targeted FISH reveals a submicroscopic deletion of chromosome 22 (Fig. 32.3a). The deletion could have been detected by aCGH (Fig. 32.3b) which also demarcates the extent of the deletion with a precision that depends on the number of DNA segment probes across the region. The aCGH hybridization results are displayed in the left track, showing the area of decreased comparative hybridization by the green dots. The center track shows the extent of deletion alongside known genes and the right track known CNVs, the genes and CNVs derived from the UCSC genome browser [10]. The deletion spans base pairs 17,299,742–19,835,558 with a minimum size of 2.54 Mb. Reference to the Decipher website [10] would confirm that the 22q11 deletion has been commonly reported and correlates with phenotypes including the DiGeorge

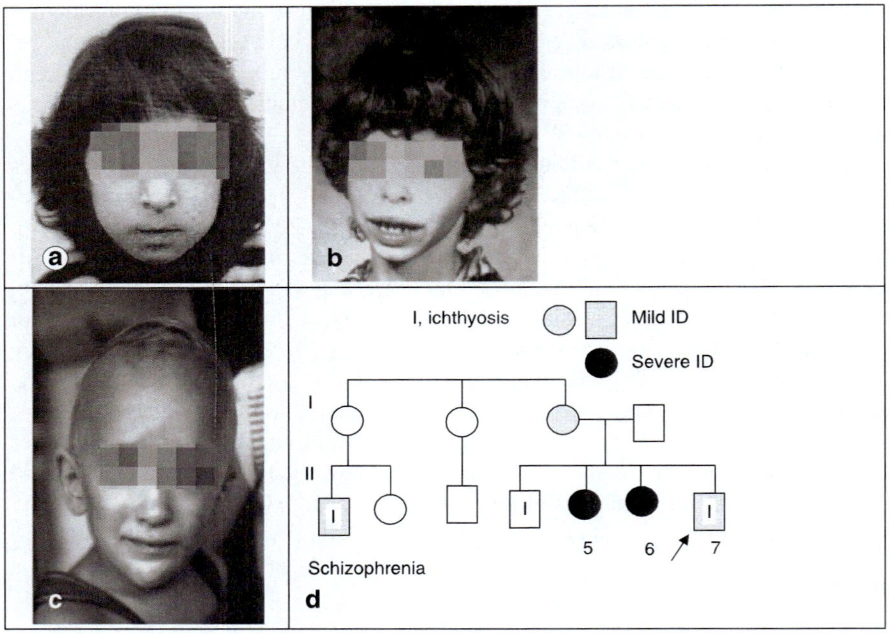

Fig. 32.2 **a** Case 1; **b** Case 2; **c** Case 3; **d** pedigree for family in Case 6

anomaly and Shprintzen/velocardiofacial syndrome. The size of deletion and the many affected patients provide confirmation of its clinical significance despite the presence of many CNVs in that region.

The size of the 22q11 microdeletion appears not to vary among patients or within families. There are ~24 genes in the deletion region (see last panel of Fig. 32.3b), with TBX1 being plausibly related to the developmental defects (heart, branchial arches) based on mouse knock-outs [11]. Relating the autistic and schizophrenic behaviors of many patients to particular genes is more difficult, with the catechol-O-methyltransferase (COMT) and GNB1L (G-protein beta-subunit like) genes being implicated [12]. The variability of findings in the 22q11 microdeletion spectrum, plus the difficulties of relating any single gene to human behaviors, introduces a recurring theme in aCGH case studies: predicting cognitive function and behavior differences is difficult even with common microdeletions/microduplications and very challenging with rare or inherited changes. At present, there seems to be no prognostic value in quantifying 22q11 deletion extent and the genes involved by aCGH; perhaps future association of particular SNPs in the deleted genes with behavior differences and cognitive function will provide an advantage for SNP array analysis over the standard FISH test.

Another question regarding the use of standard karyotype with targeted FISH versus aCGH as a first-tier test concerns which chromosome disorders should be considered "obvious" as described in the consensus statement [9]—mild cases of deletion 22q11 may not be recognized by pediatricians or less experienced clinical geneticists. The usual cost of routine chromosome studies including a single FISH analysis is \$ 600–650, that of aCGH \$ 1,500–3,000 depending on payer. Regardless of the test chosen for the child, examination of the 7% chance that a parent has the 22q13 deletion [12] can be performed using the less expensive FISH test. Choices between standard karyotype with targeted FISH, FISH panels covering the more common submicroscopic deletions (e. g., for 1p36 deletion, Williams, Prader-Willi, velocardiofacial syndromes), and aCGH as first-line tests will depend on the

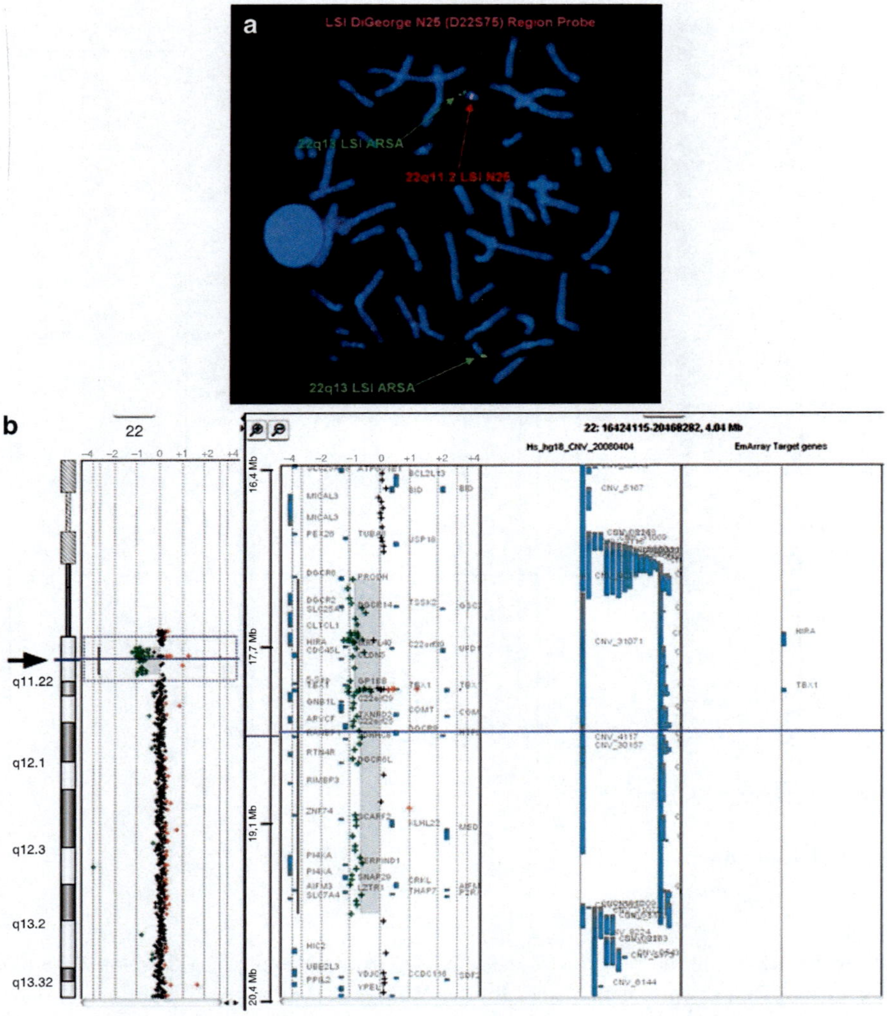

Fig. 32.3 **a** Case 1 FISH analysis showing absent test signal on one chromosome 22; **b** Case 1 aCGH analysis showing deletion of 2.54 Mb at band 22q11.21; **c** Case 2 aCGH analysis showing deletion of 1.37 Mb at band 7q11.23; **d** Case 3 aCGH analysis showing deletion of 1.86 Mb at band 5q35.2q35.3. Large arrow at far left of each array points to heavy blue line on idiogram and to boxed area (defined by blue dotted line) on array histogram showing deletion for each case

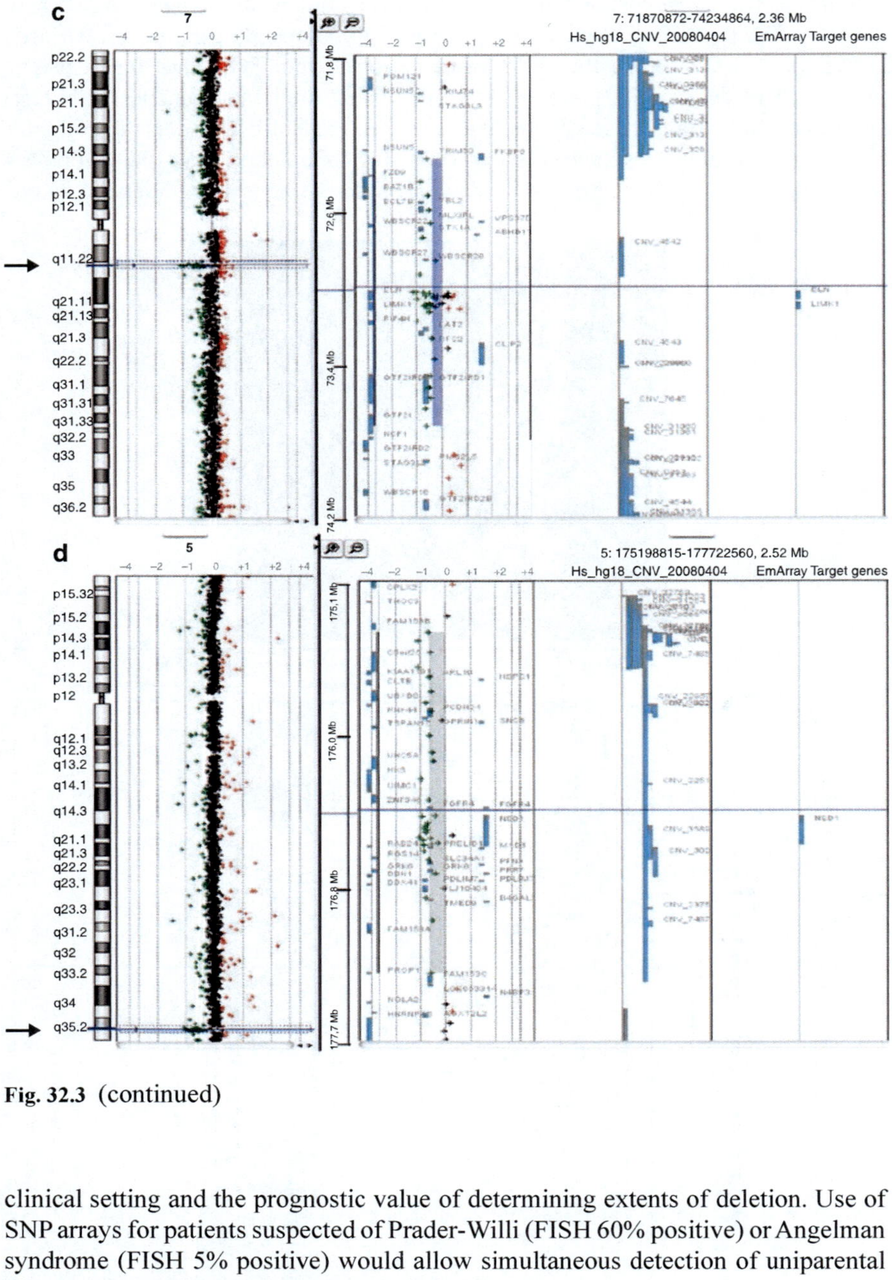

Fig. 32.3 (continued)

clinical setting and the prognostic value of determining extents of deletion. Use of SNP arrays for patients suspected of Prader-Willi (FISH 60% positive) or Angelman syndrome (FISH 5% positive) would allow simultaneous detection of uniparental disomy but not abnormal DNA methylation cases.

Case 2: A 4-year-old girl (Fig. 32.2b) had a normal gestation and family history but developed feeding problems in the nursery. She continued to have colic with multiple formula changes and was evaluated for failure to thrive at age 10 months with no definitive diagnosis. Referral to pediatric genetics revealed a stature and weight at the 3rd centile for age with head size below the 3rd centile. She had coarsened facial appearance with thick lips (Fig. 32.2b), increased joint laxity, and a language level of a 2-year-old. She had some hyperactive behaviors and a remarkable happy affect.

The aCGH study (Fig. 32.3c) documented deletion of 1.37 Mb at band 7q11.23, its size and prior correlation with numerous Williams syndrome patients establishing its significance. Among different affected patients, the deletion size varies somewhat (1.5–1.8 Mb per reference 13) but does not correlate with clinical severity—thus standard FISH analysis is as useful as aCGH if clinical recognition targets the appropriate FISH test. The ~28 haplo-insufficient genes include *ELN* (elastin) that mediates connective tissue signs like joint flexibility and cardiovascular disease. The unusual behavioral phenotypes of happy affect and musical ability in Williams syndrome, like other cognitive functions, are likely related to multifactorial determination, parental origins, and gene interactions rather than to any single gene within the deletion [13]. There is a reciprocal microduplication, usually with more severe speech delays, arising from low copy repeats that border the deletion/duplication region. As with Shprintzen-DiGeorge spectrum, Williams syndrome exhibits dramatic clinical variability, with most patients having borderline mental disability (IQ 70–80) and others having autism disorder.

Case 3: A 4-year-old boy (Fig. 32.2c) had a birth weight of 9 lbs 6 oz after a non-diabetic pregnancy and had trouble breast-feeding because of hypotonia. Early growth was rapid with the head size accelerating above the 97th centile, necessitating head MRI study that showed normal brain structure with mild ventricular dilation. The teeth came in early, indicating an accelerated bone age, but the motor milestones were delayed due to hypotonia with walking at age 2 years. Pediatric genetic evaluation showed a prominent forehead and triangular facial shape with down-slanting palpebral fissures (Fig. 32.2c).

The aCGH analysis (Fig. 32.3d) shows a 1.86 Mb microdeletion at band 5q35.2q35.3 that is typical of patients with Sotos syndrome (cerebral gigantism). Microduplications or microdeletions of this region are detected in 50% of Japanese and 10–15% of Occidental patients, with most of the remainder having mutations within the *NSD1* gene that affects histone processing [14].Use of aCGH provides slightly greater detection rates (up to 15%) versus 10% for standard FISH testing in Western populations [14]. The corresponding *NSD1* mutations focus attention on this gene among the several in the duplication/deletion area, but here again limited knowledge of gene function hampers correlation with the variable mental disability and growth alterations of the Sotos phenotype. Complicating these correlations are the 5% of Sotos patients who have an affected parent: although the mutations seem fully penetrant [14], parents may have such mild features that they have never presented for medical evaluation.

32.2.2 Cases Where Interpretation of Clinical Significance Is Clear but the Finding Gives Less Defined Prognosis

Case 4: A 5-year-old boy presents for evaluation of genetic causes of autism disorder. He had normal motor development but presented for developmental evaluation because of absent speech at age 2.5 years. He had exhibited hypersensitivity to loud or background noises and made poor eye contact with preference for adult interactions rather than those with other children. Physical examination was entirely normal with normal growth and morphology, and he had no neurologic deficits. He did exhibit some repetitive movements with hand-flapping and finger-motions and was fascinated with objects such as venetian blinds, light switches, and toy car wheels. He would not interact with the examiner and exhibited no reciprocal communication. The parents desired genetic testing to define risks for future pregnancies and for the patient's healthy sister.

Routine chromosome and fragile X DNA analysis were performed and were normal; reflex to aCGH analysis revealed a 1.43 Mb duplication at chromosome band 16p13.12p13.11 (Fig. 32.4a), close to the 16p11 duplications or deletions found in 2–3% of patients with autism collected by an international consortium [15]. Parental studies were normal. This result and those in other patients with autism and/or intellectual disability without significant dysmorphology demonstrate the increased yield of aCGH genetic testing, increasing the frequency of positive findings to 10–15%, a number 3 to 5 times greater than that for routine chromosome analysis. Note several genes and numerous CNVs across the 16p duplication interval (right tracks, Fig. 32.4a), again making the correlation with duplication/deletion and mental function uncertain. This patient with significant cognitive/speech delay would qualify clinically as autism disorder, but minimal clinical description of patients in consortia does not allow correlation of particular deletions with autism/ID severity. However, the aCGH finding does establish a cause for the disorder, alleviates maternal guilt about pregnancy behaviors being at fault, and (assuming normal parental studies), specifies a low recurrence risk. Since autistic symptoms can occur in any disorder with mental disability, from Down to Prader-Willi syndromes, the aCGH finding correlates with but does not specify a diagnosis of autism.

Case 5: A 3-year-old girl was evaluated because of speech delay and autistic symptoms. She had sensory signs, poor eye contact, early feeding problems, and signifi-

Fig. 32.4 **a** Case 4 aCGH analysis showing 1.43 Mb duplication at band 16p13.12p13.11; **b** Case 5 aCGH analysis showing deletion of 3.26 Mb at band 15q13.1q13.3 in the proband (track 1) and her brother (track 2); **c** Case 6 aCGH showing 1.56 Mb deletion at band Xp22.31 for the proband (II-5) and two sisters (II-6, II-7 in Fig. 2D); **d** Case 7 aCGH showing 0.35 Mb duplication (track 2) at band 9p13.3 along with positive paternal (track 1) and negative maternal results (track 3). Large arrow at far left of each array points to heavy blue line on idiogram and to boxed area (defined by blue dotted line) on array histogram showing deletion or duplication for each case. Large arrow at far left of each array points to heavy blue line on idiogram and to boxed area (defined by blue dotted line) on array histogram showing deletion or duplication for each case

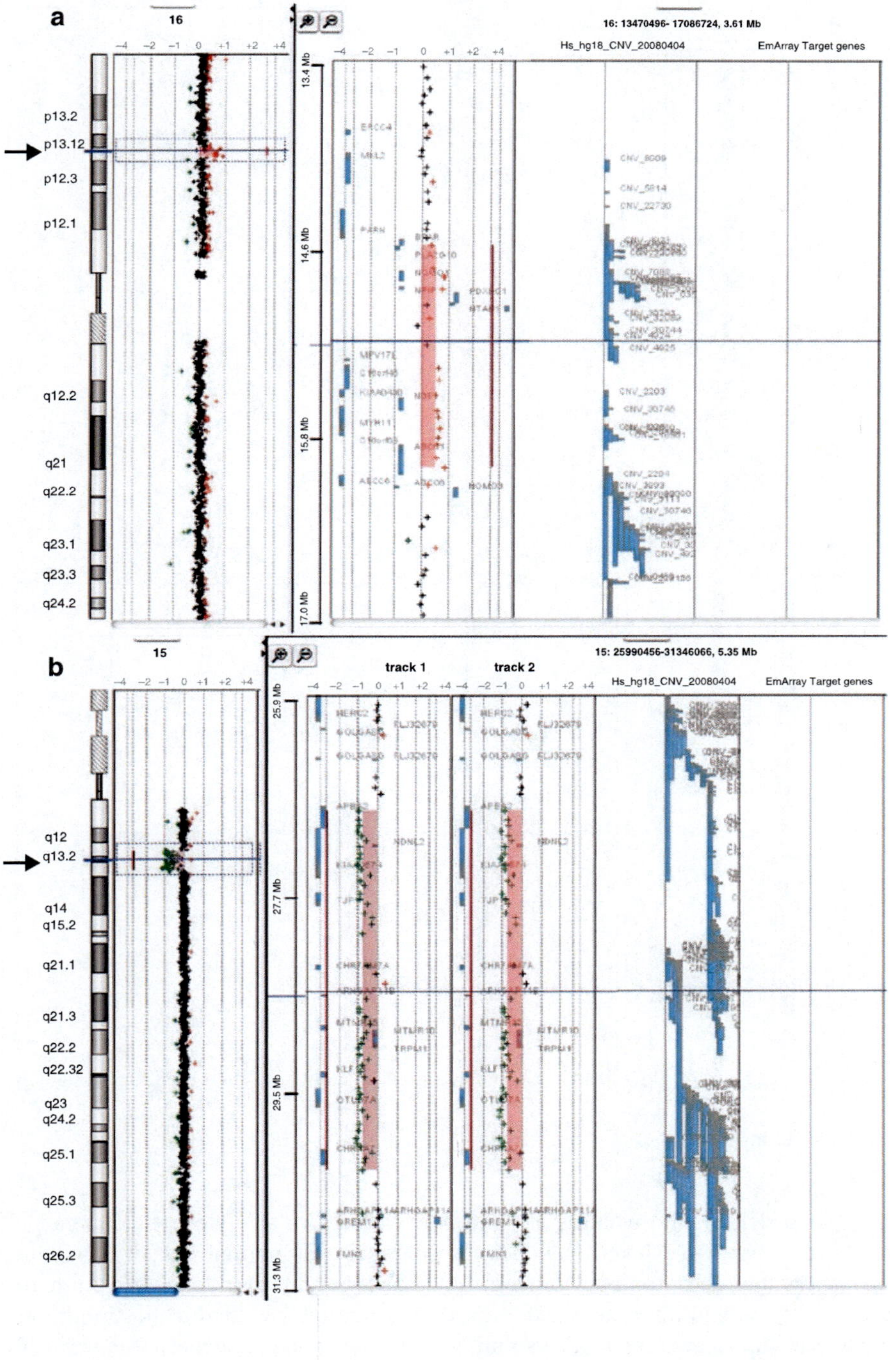
a
16
p13.2
p13.12
p12.3
p12.1
q12.2
q21
q22.2
q23.1
q23.3
q24.2
16: 13470496- 17086724, 3.61 Mb
Hs_hg18_CNV_20080404
EmArray Target genes
13.4 Mb
14.6 Mb
15.8 Mb
17.0 Mb
b
15
q12
q13.2
q14
q15.2
q21.1
q21.3
q22.2
q22.32
q23
q24.2
q25.1
q25.3
q26.2
15: 25990456-31346066, 5.35 Mb
track 1
track 2
Hs_hg18_CNV_20080404
EmArray Target genes
25.9 Mb
27.7 Mb
29.5 Mb
31.3 Mb

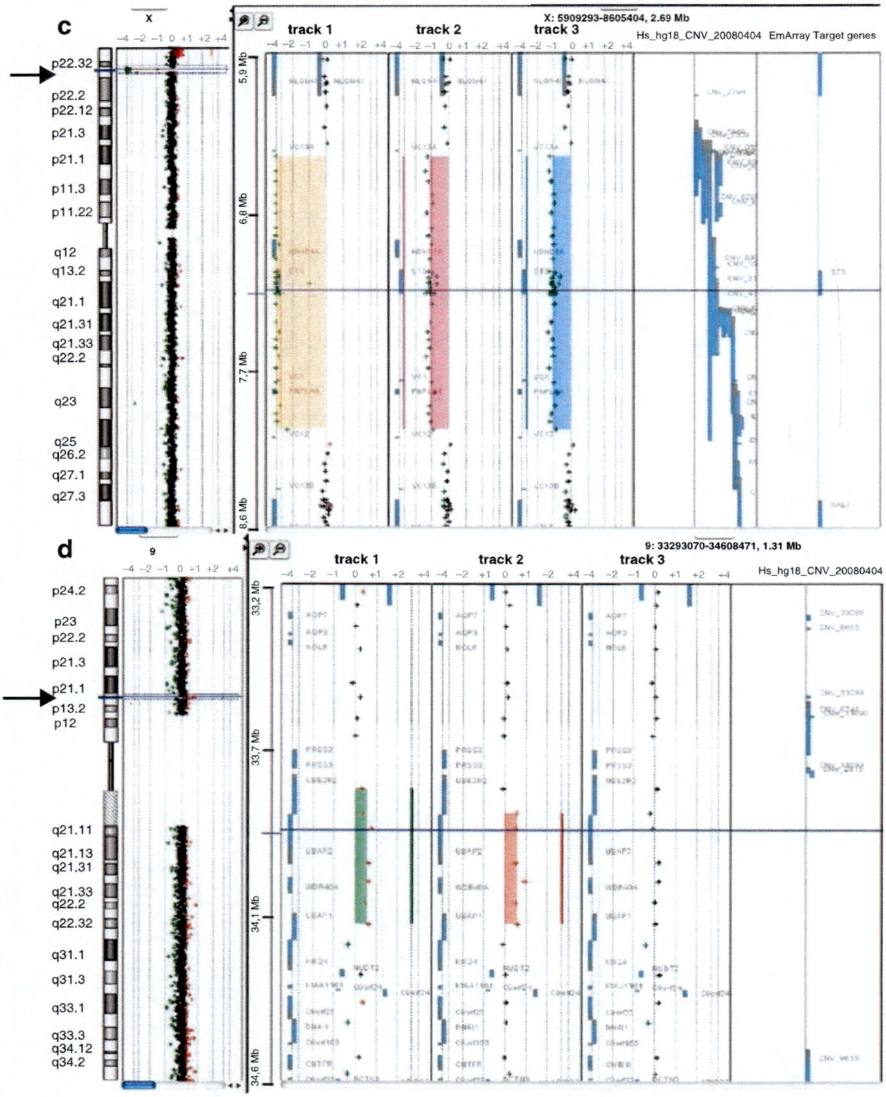

Fig. 32.4 (Continued)

cant motor delays with walking at 18 months. Receptive and expressive language was at the 2-year-old level. Her facial appearance was normal except for slight coarseness that had prompted concern for a mucopolysaccharidosis; urine mucopolysaccharide and oligosaccharide testing was normal. She had a short philtrum and everted lip resembling that of Smith-Magenis syndrome. She had a 9-month-old brother with seizures and motor delays.

Initial routine chromosome analysis was normal with reflex to aCGH analysis demonstrating a 2.26 Mb deletion at band 15q13.1q13.3 in the proband (track 1) and her brother (track 2—Fig. 32.4b); parental studies showed a similar deletion in the father although he had few medical symptoms aside from learning differences that led him to leave high school before graduation. Although parental studies have been positive in the majority of reported cases [16], the presence of identical 15q13.3 microdeletions in over 50 patients with mental disability and/or behavior differences argues for clinical significance. The variable expressivity [16] again confounds the correlation of frequent autism, mental disability, or mental illness with the many genes within the 15q13.3 deletion interval (Fig. 32.4b, left tracks). The problem of clinical correlation and prognosis is exemplified further in the families below, especially in the last case where father and son have the same 9p deletion that is atypical for the son's clinical diagnosis.

32.2.3 Cases of Familial Change with Unclear Significance or Prognosis

Case 6: A 9-year-old boy (individual II-7 in Fig. 32.2d) presented for genetic evaluation because of large size, school problems, familial ichthyosis, and possible autism. He had early motor delays with mild hypotonia and later speech delays requiring therapy. He did poorly in school and was found to have a borderline IQ of 85 with autistic tendencies. Prior routine chromosome analysis had shown an Xp22.31 deletion and DNA testing for the neuroligin 4X gene near this region had been normal. Family history was significant (Fig. 32.2d) for two sisters with mental disability, a normal brother with ichthyosis, a mother with anxiety disorder, maternal brothers with ichthyosis, and offspring of mother's sisters with mental disability or schizophrenia. The boy's physical examination was normal, and aCGH analysis was performed on the family including his mother, two sisters, and healthy brother.

aCGH analysis demonstrated an identical 1.56 Mb deletion at band Xp22.31 in the mother and her four children—analyses of the sisters II-5, II-6 and proband II-7 are shown in Fig. 32.4c. Extreme variability of expression with deletions in the Xp22.31 region has been documented in many reports [17], illustrated in this family by normality in one brother, mild mental disability in the patient, more severe mental disability in his two sisters, variable autistic features, and anxiety disorder in the mother. Thus behavior changes range from autism to anxiety disorder to schizophrenia in a nephew who based on his ichthyosis likely has the same Xp22 deletion. There are numerous genes in the deletion interval, but correlations are again hampered by limited knowledge of their function. The VCX gene cluster, with potential for non-homologous gene recombination, is the subject of current studies [17].

Case 7: A 5-year-old boy presented for evaluation of developmental delays and had the characteristic facial appearance of Rubinstein-Taybi syndrome. He was found to have a 0.35 Mb duplication (track 2) at band 9p13.3 that was also present in his father (track 1) but not his mother (track 3, Fig. 32.4d). His father had had no learning differences or physical anomalies and his family history was normal. The 9p duplication has not been previously associated with Rubinstein-Taybi syndrome where about 10% of patients will have a deletion of 16p encompassing the CREBBP transcription factor that can be detected by FISH [18]. Another 30–50% of affected individuals will have mutations in the CREBBP gene, while 3% will have mutations in the EP300 gene (encoding a histone acetyltransferase) at 22q13. Although the deletion is above the 0.5 Mb size generally considered significant, its presence in the normal father and lack of correlation with the child's phenotype makes interpretation very difficult. Continued accumulation of CNVs in computerized databases [10] will improve interpretation provided that sufficient clinical data are included.

32.3 Summary

The progression from routine chromosome analysis to high resolution aCGH has greatly increased the sensitivity of genetic testing for children with intellectual disability and behavior differences, extending diagnostic findings to those without obvious dysmorphology. This increased sensitivity has brought with it challenges exemplified by our sample cases, showing that the combined criteria of size, inheritance, gene content, or population frequency are not always sufficient to determine clinical significance. Continued accrual of patient profiles and CNV findings in several databases [10], along with studies of gene expression/interaction and epigenesis, should improve future clinical correlations. The technique of aCGH is an important addition to the genetic testing repertoire, but pre-test counseling regarding difficulties of interpretation and lack of prognostic or therapeutic guidance is important when discussing the technique with parents. Use of more expensive aCGH as a first-tier test [9] must be balanced against the lack of detection of balanced rearrangements and the likelihood that clinicians can specify the appropriate FISH testing. These problems of interpretation and economy will only become more challenging if aCGH is recommended for use in routine prenatal diagnosis.

References

1. Lee C, Iafrate A, Brothman AR (2007) Copy number variations and clinical cytogenetic diagnosis of constitutional disorders. Nature Genet suppl 39:S48–S54
2. Carter NP (2007) Methods and strategies for analyzing copy number variation using DNA microarrays. Nature Genet suppl 39:S16–S21
3. Feuk L, Marshall CR, Wintle RF, Scherer SW (2006) Structural variants: changing the landscape of chromosomes and the design of disease studies. Hum Molec Genet 15:R57–R66

4. Lupski JR (1998) Genomic disorders: structural features of the genome can lead to DNA rearrangements and human disease traits. Trends Genet 14:417–422
5. Sebat J, Lakshmi B, Troge J et al (2004) Large—scale copy number polymorphisms in the human genome. Science 305:525–528
6. Freeman JL, Perry GH, Feuk L et al (2006) Copy number variation: new insights in genome diversity. Genome Research 16:949–961
7. Kaiser-Rogers K, Rao K (2009) Structural chromosome rearrangements. In: Gersen SL, Keagle MB (eds) Principles of clinical cytogenetics, 2nd edn. Human, Totowa, pp 165–206
8. Sagoo GS, Butterworth AS, Sanderson S, Shaw-Smith C, Higgins JPT, Burton H (2009) Array CGH in patients with learning disability (mental retardation) and congenital anomalies: updated systematic review and meta-analysis of 19 studies and 13926 subjects. Genet Med 11:139–146
9. Miller DT, Adam MP, Aradhya S et al (2010) Consensus Statement: Chromosomal microarray is a first-tier clinical diagnostic test for individuals with developmental disabilities or congenital anomalies. Am J Hum Genet 86:749–764
10. Web Resources: http://cibex.nig.ac.jp/index.jsp, UC Santa Clara genome browser: http://genome.ucsc.edu/, Decipher: https://decipher.sanger.ac.uk/application/, http://projects.tcag.ca/variation/, http://www.ncbi.nlm.nih.gov/projects/SNP/, http://humanparalogy.gs.washington.edu/structuralvariation/, http://uswest.ensembl.org/index.html, http://www.ncbi.nlm.nih.gov/geo/, http://www.genenames.org/ http://humanparalogy.gs.washington.edu/; http://ccr.coriell.org/Sections/Collections/NIGMS/?SsId=; http://projects.tcag.ca/humandup/, http://www.ncbi.nlm.nih.gov/unigene
11. Scambler PJ (2010) 22q11 deletion syndrome: a role for TBX1 in pharyngeal and cardiovascular development. Pediatr Cardiol 31:373–390
12. Gothelf D, Schaer M, Eliez S (2008) Genes, brain development and psychiatric phenotypes in velo-cardio-facial syndrome. Dev Disabil Res Rev 14:59–68
13. Brunetti-Pierri MG, Micale L, Fusco C (2010) Copy number variants at Williams-Beuren syndrome 7q11.23 region. Hum Genet 128:3–26
14. Tatton-Brown K, Cole TRP, Rahman N (2009) Sotos syndrome In: Pagon RA, Bird TC, Dolan CR, Stephens K, eds. Gene Reviews [Internet]. Seattle (WA): University of Washington, Seattle; 1993–2004 Dec 17 (Updated 2009 Dec 10)
15. Weiss LA, Shen Y, Korn JM et al (2008) Association between microdeletion and microduplication at 16p11.2 and autism. N Engl J Med 358:667–675
16. Ben-Schachar S, Lanpher B, German JR et al (2009) Microdeletion 15q13.3: a locus with incomplete penetrance for autism, mental retardation, and psychiatric disorders. J Med Genet 46:382–88
17. Van Esch H, Hollanders K, Badisco L, Melotte C, Van Hummelen P, Vermeesch JR, Devriendt K, Fryns JP, Marynen P, Froyen G (2005) Deletion of VCX-A due to NAHR plays a major role in the occurrence of mental retardation in patients with X-linked ichthyosis. Hum Mol Genet 14:1795–1803
18. Stevens CA (2009) Rubinstein-Taybe syndrome In: Pagon RA, Bird TC, Dolan CR, Stephens K, (eds) Gene Reviews [Internet]. Seattle (WA): University of Washington, Seattle; 1993–2002 Aug 30 (Updated 2009 Aug 20)

Index

H. E. Wyandt, V. S. Tonk, *Human Chromosome Variation: Heteromorphism and Polymorphism,* DOI 10.1007/978-94-007-0896-9,